Spring Meeting Poster Session and Networking Reception 2017

Core Programming Area at the 2017 AIChE Spring Meeting and 13th Global Congress on Process Safety

San Antonio, Texas, USA
26 – 30 March 2017

ISBN: 978-1-5108-4150-5

Printed from e-media with permission by:

Curran Associates, Inc.
57 Morehouse Lane
Red Hook, NY 12571

Some format issues inherent in the e-media version may also appear in this print version.

Printed by Curran Associates, Inc. (2017)

For permission requests, please contact AIChE
at the address below.

AIChE
120 Wall Street, FL 23
New York, NY 10005-4020

Phone: (800) 242-4363
Fax: (203) 775-5177

www.aiche.org

Additional copies of this publication are available from:

Curran Associates, Inc.
57 Morehouse Lane
Red Hook, NY 12571 USA
Phone: 845-758-0400
Fax: 845-758-2633
Email: curran@proceedings.com
Web: www.proceedings.com

TABLE OF CONTENTS

478213 Process Safety Problems Caused By Hydrate Formation in Deepwater Production Operations

Abstract:

Abstract for 2017 AIChE Spring Meeting and 13th Global Congress on Process Safety March 26-30, 2017 San Antonio, TX Session Selection: 19th Process Plant Safety Symposium (PPSS) Using Date form Incident Investigation, Risk Assessments, and Mechanical Integrity to improve Process Safety Performance.

Process Safety Problems Caused by Hydrate Formation in Deepwater Production Operations

Susana Leon, Yogesh Koirala, Sam Mannan

Mary Kay O'Connor Process Safety Center

Artie McFerrin Department of Chemical Engineering

Texas A&M University

College Station, Texas 77843-3122, USA

+1(979)845-3489, Susana_leon@tamu.edu

Abstract: Hydrate formation has been seen as one of the most challenging problems in deepwater production operations because of the low temperatures and high pressures found in deepwater environments. Hydrate formation and its agglomeration in subsea production equipment can cause severe operational issues, leading to substantial asset damage and, most importantly, can cause significant process safety and environmental issues. Examples of the problems caused by hydrates include blockage in wells, risers, flowlines, and manifolds. These blockages can increase the pressure in the system causing equipment damage and potentially resulting in the loss of hydrocarbon containment into the environment.

Although hydrate formation research has been conducted for many years, it has been focused in the development of conceptual models to understand these phenomena and to determine methods to prevent and control the risk. However, recently, the new research approach is to develop a risk assessment in order to have an accurate prediction of hydrate formation in different oil and gas processes that operate under a wide range of conditions. Thus, it is extremely necessary to have a range of probabilities of hydrate formation rather than a fixed and single point, which will allow for the implementation of better strategies to manage the risk associated whit this phenomena. Therefore, this research is focused in the revision, and evaluation of deterministic and probabilistic methods looking for opportunities to optimize the prediction of hydrate formation in subsea production areas; for example, application of the Bayesian probability theory. After having a better estimate of the probability, this research intends to explore the risk assessment methodologies and apply them to subsea processes as part of a new strategy plan to prevent and mitigate the risk.

With the purpose of developing a risk assessment approach to quantify the risk related to the hydrate formation, it is necessary to identify the best hydrate formation model to represent the phenomena behavior followed by the utilization of the Bayesian probability theory to optimize the prediction of hydrate formation. The last step in the process is to identify possible consequences of hydrate formation issues at subsea area and implement an accurate methodology to quantify the risk using hydrate formation models.

Existing literature review shows that the oil and gas industry is taking a different approach to hydrate formation issues, including risk quantification and risk assessment approach.

A risk quantification approach is relatively a new research topic. Therefore, research institutions together with industry started to go deep into how to successfully manage the hydrate formation risk and have found that it requires an accurate forecast of the hydrate growth rate. Consequently, this research will lead to a better understanding of the risk quantification.

Keywords: Safety, hydrate, Bayesian networks, risk assessment

"Photo Catalytic Degradation Of Pharmaceutical Personal Care Products (PPCP) In Wastewater Using Rhodamine B Dye As A Surrogate".

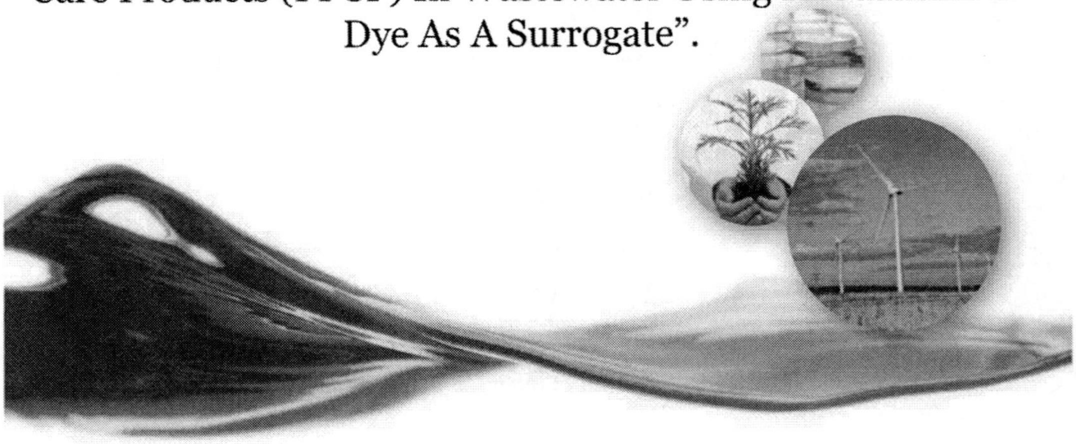

AIChE 2017 Spring Meeting and 13th GCPS

AUTHOR & CO-AUTHORS

Abibat Ahmed

Hassan Alsaud

Dr. Roger Painter

Dr. Lonnie Sharpe

Dr. Samuel Hargrove

Civil Engineering, Tennessee State University

Acknowledgement

Project Overview

- Introduction
- Literature Review
- Materials and Methods
- Results
- Discussion
- Conclusion and Recommendation

Introduction

Significance of the research

Objective and scope of the investigation

Method of investigation

Organization of report

Literature Review

- Wastewater pollution
- Contaminants of emerging concern
  e.g. industrial chemicals, pharmaceutical personal care products (PPCP), disinfectants by-products, surfactants, textile dyes etc.
- PPCP toxicity to the environment
- Rhodamine B dye: a surrogate for PPCP
- Advanced oxidation treatment process
  e.g. photo catalysis, UV radiation, Photo-fenton etc.

- Photocatalytic degradation of Rhodamine B dye presence of metal ions, pH, temperature, surfactants etc affects the rate of degradation.

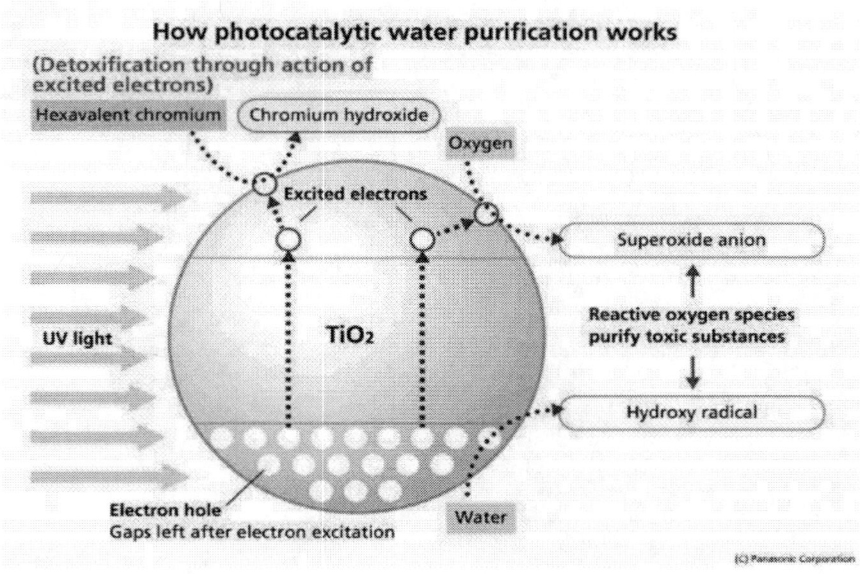

Reaction Kinetics

- Holes and electrons are generated

$$TiO_2 \xrightarrow{ke} h^+_{VB} + e^-$$

- Hydroxyl radicals are produced

By holes:

$$S\text{—}OH^- + h^+_{VB} \Leftrightarrow S\text{—}OH^\cdot$$

$$S\text{—}H_2O + h^+_{VB} \Leftrightarrow S\text{—}OH^\cdot + H^+$$

By electrons

$$e^- + O_2 \xrightarrow{Ke1} O_2^{\cdot-}$$

$$O_2^{\cdot-} + e^- + 2H^+ \xrightarrow{Ke2} H_2O_2$$

$$2O_2^{\cdot-} + 2H^+ \xrightarrow{Ke3} H_2O_2 + O_2$$

$$H_2O_2 + e^- \xrightarrow{Ke4} OH^\cdot + OH^-$$

Reaction Kinetics

- Dye Molecules are adsorbed by catalyst

$$S + D \overset{k_D}{\leftrightarrow} S—D,\ K_D = k_D / k_D'$$

- Dye molecules (adsorbed or free) are mineralized

By holes:

$$S—D + h_{VB}^{+} \xrightarrow{Khr,a} S—D^{\circ +} \rightarrow \text{end species}$$

$$D + h_{VB}^{+} \xrightarrow{Khr,b} D^{\circ +} \rightarrow \text{end species}$$

By OH·

$$S—D + S—OH\cdot \xrightarrow{Kra} S + S—D'^{0} \rightarrow \text{end species}$$

$$D + S—OH\cdot \xrightarrow{Krb} S + D'^{0} \rightarrow \text{end species}$$

$$S—D + OH\cdot \xrightarrow{Krc} S + S—D'^{0} \rightarrow \text{end species}$$

$$D + OH\cdot \xrightarrow{Krd} S + D'^{0} \rightarrow \text{end species}$$

Reaction Kinetics

$$-r_d = k_{hr,a}[S\text{-}D]\ [h_{VB}^{+}] + k_{hr,b}[D]\ [h_{VB}^{+}] + k_{ra}[S\text{-}D][S\text{-}OH\cdot] + k_{rb}[D][S\text{-}OH\cdot] + k_{rc}[S\text{-}D][OH\cdot] + k_{rd}[D][OH\cdot]$$

$$-r_D = k_{oh}\ [D] + k_1\ [D]\ \{\frac{k_3}{1+K_2+[D]} + \{\frac{k_4}{1+K_2[D]} * \frac{1}{1+K_6[Cu^{2+}]}\}\}$$

$$\frac{1}{r_D} = \left(\frac{1}{[D]} + K_2\right)\frac{(1+K_6[Cu^{2+}])}{(k_0 + K_6 k_7 [Cu^{2+}])}$$

$$\frac{1}{r_D} = \left(\frac{1}{[D]} + K_2\right)\frac{1}{k_0}$$

Where:

$$k_0 = k_{oh} + k_1 k_3 + k_1 k_4$$

$$k_7 = k_{oh} + k_1 k_3$$

Materials and Method

Apparatus

- Photochemical Reactor
  - Magnetic Stirrer
  - 2-liter flask
  - LED stripe 380 nm
  - Power source
  - 10ml pipette
  - Test tube

Analytical weigh balance

Fluorimeter

Materials and Methods

- Rhodamine B → 500 ppb, 400 ppb, 300 ppb, 200 ppb and 100 ppb
- Copper ion solution → 500 ppb, 400 ppb, 300 ppb, 200 ppb and 100 ppb
- Titanium dioxide (TiO_2) → 0.05 grams
- De-ionized water
- Sampling every hour
- Fluorescence by comparing light.

Results and Discussion

500 µg/L Rhodamine B dye 0.05g/L TIO2						
	Initial Copper Ion concentration					
	500 µg/L	400 µg/L	300 µg/L	200 µg/L	100 µg/L	0 µg/L
Time	Dye Absorbed					
0	500.0	493.0	480.0	478.0	472.0	467.0
60	464.3	420.0	382.0	369.0	352.5	317.8
120	355.0	335.2	281.0	257.4	212.3	235.0
180	299.0	271.0	198.0	173.0	111.4	161.0
240	263.1	220.0	169.0	121.2	76.1	87.0
300	162.3	186.0	123.7	81.0		45.0

Results and Discussion

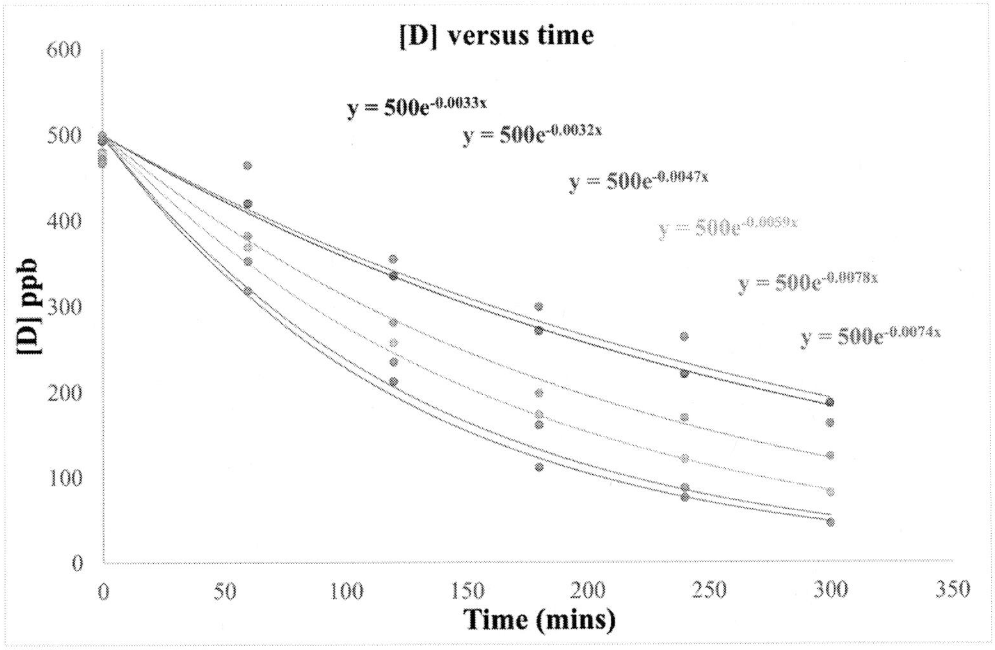

Results and Discussion

0 µg/L Copper concentration 0.05g/L TIO2						
	Rhodamine B initial concentration					
	500 µg/L	400 µg/L	300 µg/L	200 µg/L	100 µg/L	375 µg/L
Time	**Dye Absorbed**					
0	500	393	297	206	103	361
60	382	252.6	209	128.4	69	236
120	235	187.6	98	69.2	38	126
180	161	101	40	42.1	-	67
240	87	81.9				

Results and Discussion

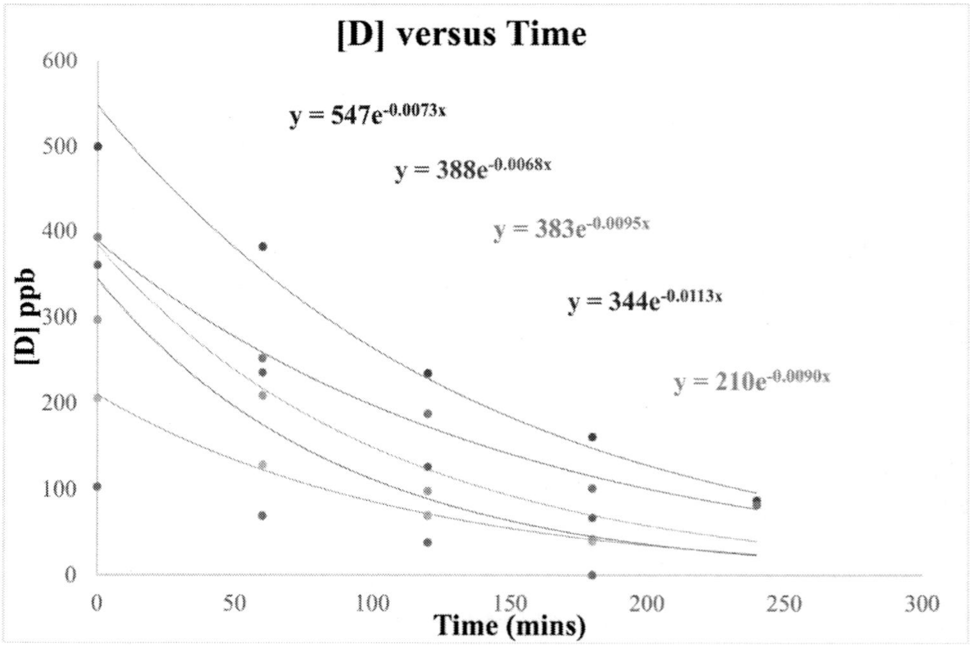

Linear Regression

t (mins)	R(ug/l/min)	R (mg/l/min)	1/r	[D] (ug/l)	[D] (mg/l)	1/[D]
0	3.700	0.004	270.27	467.0	0.467	2.14
60	2.373	0.002	421.33	317.8	0.318	3.15
120	1.522	0.002	656.83	235.0	0.235	4.26
180	0.977	0.001	1023.95	161.0	0.161	6.21
240	0.626	0.001	1596.27	87.0	0.087	11.49
300	0.402	0.000	2488.47	45.0	0.045	22.22

Linear Regression

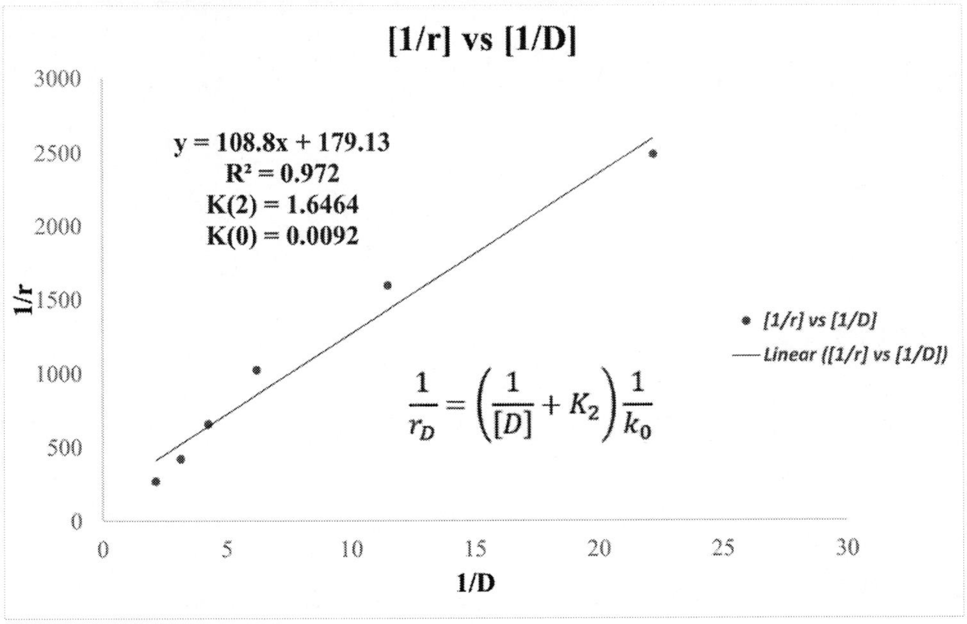

Linear Regression

Regression Statistics		correlation	k(0)	k(2)
Multiple R	0.9872	0.99	0.0092	1.6464
R Square	0.9745			
Adjusted R Square	0.9660			
Standard Error	152.6302			
Observations	5.0000			

ANOVA

	df	*SS*	*MS*	*F*	*Significance F*
Regression	1	2672658.4037	2672658.4037	114.7262	0.0017
Residual	3	69887.9037	23295.9679		
Total	4	2742546.3074			

Non-linear Regression

K	r(ug/l/min)	r(mg/l/min)	1/r	[Cu2+] (mg/l)
0.0074	3.7	0.0037	270.27	0.00
0.0078	3.90	0.0039	256.41	0.10
0.0059	2.95	0.0030	338.98	0.20
0.0047	2.35	0.0024	425.53	0.30
0.0032	1.60	0.0016	625.00	0.40
0.0033	1.65	0.0017	606.06	0.50

Non-linear Regression

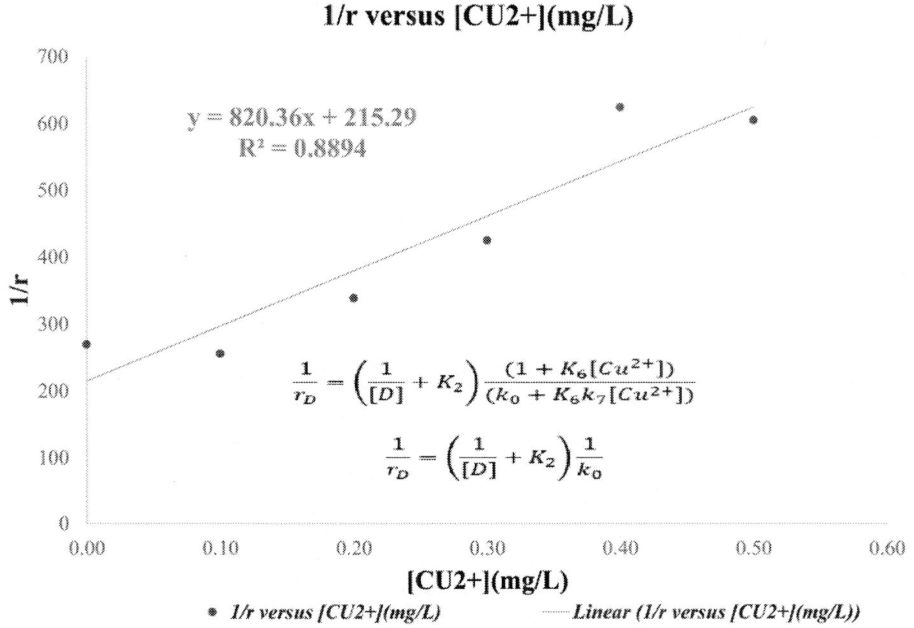

Degradation Pathway

$$-r_D = k_{oh}[D] + k_l [D] \left\{\frac{k_3}{1+K_2+[D]} + \left\{\frac{k_4}{1+K_2[D]} * \frac{1}{1+K_6[Cu^{2+}]}\right\}\right\}$$

Direct Hole Attack

$$S—D + h\nu_B^+ \xrightarrow{Khr,a} S—D^{\cdot +} \rightarrow \text{end species}$$

$$D + h\nu_B^+ \xrightarrow{Khr,b} D^{\cdot +} \rightarrow \text{end species}$$

Degradation Pathway

$$k_{oh} = \{ k_{hr,a} K_D [S] + k_{hr,b} \} [h^+_{VB}]$$

Hole pathway

$$S{-}OH^- + h^+_{VB} \overset{Ktr}{\Leftrightarrow} S{-}OH^{\bullet}$$

$$S{-}H_2O + h^+_{VB} \underset{K'tr}{\Leftrightarrow} S{-}OH^{\bullet} + H^+$$

Rate Constants findings – No Copper ion

Equation of [D]/[T]	[D]	T (mins)	d[C]/d[t] - 375ppb	r[D]	Assumed K	K1k4
$106.4184e^{-0.0083x}$	100	0	-3.56	-4.14	-5.90	-6.851
$210.0329e^{-0.0090x}$	200	60	-2.01	-2.94	-3.33	-4.851
$344.9043e^{-0.0113x}$	300	90	-1.52	-1.97	-2.50	-3.251
$383.9211e^{-0.0095x}$	375	150	-0.86	-1.40	-1.41	-2.313
$388.6858e^{-0.0068x}$	400	180	-0.64	-1.24	-1.06	-2.051
$547.6676e^{-0.0073x}$	500	240	-0.36	-0.76	-0.60	-1.251

Rate Constants findings

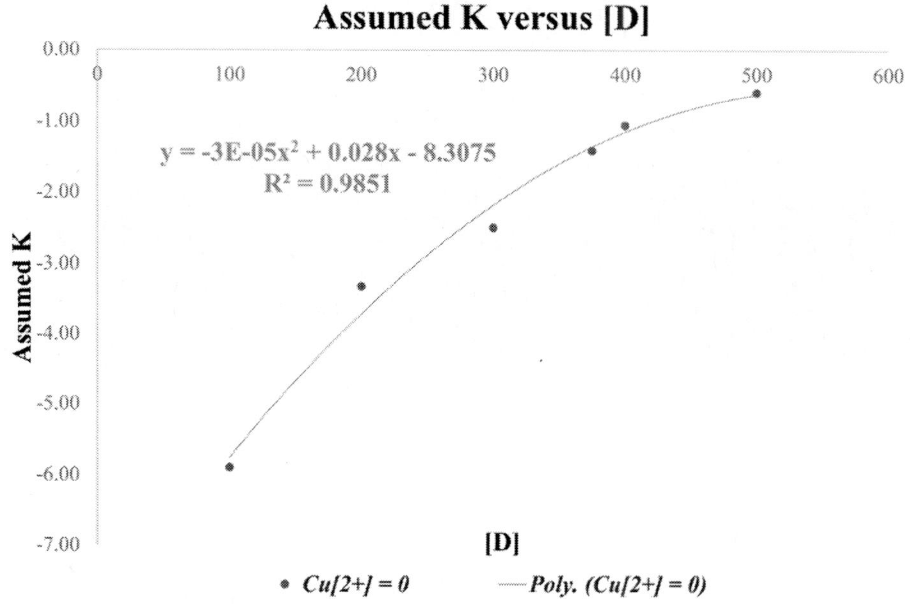

Rate Constants findings

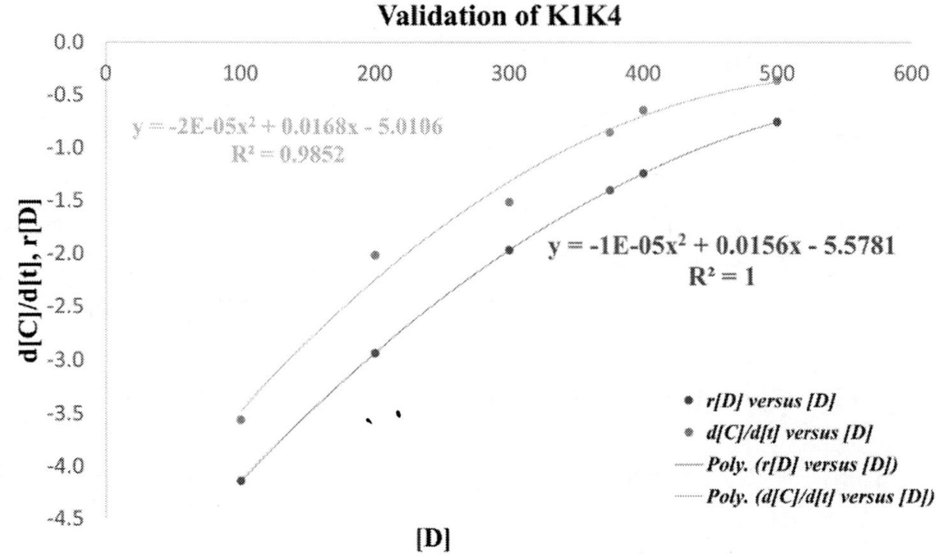

Simplified Model

$$-r_D = k_1 [D]\{\frac{k_4}{1+K_2[D]} * \frac{1}{1+K_6[Cu^{2+}]}\}$$

Adsorption

[TiO2]	[D]intial	[D]final	[S-D]	[TiO2][D]
0.00	497.50	497.50	0.00	0.00
0.25	497.50	386.13	111.37	124.38
0.50	493.30	332.80	160.50	246.65
0.75	477.70	252.33	225.37	358.28
1.00	496.80	291.07	205.73	496.80
1.50	468.20	236.40	231.80	702.30

Adsorption

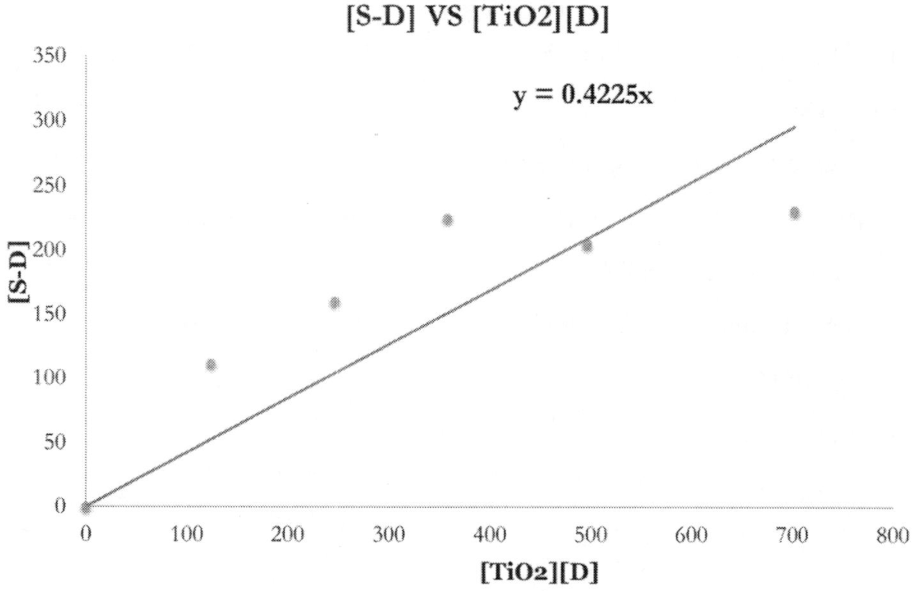

Discussion and Comparism

- Cu^{2+} ions → electrons → reduce OH. production → slows down electron pathway
- Presence of Cu^{2+} allowed isolation and differentiation of e- and hole pathway
- Hole pathway is negligible since k_7 and k_{OH} are both zero
- k_1k_4 → function of dye concentration
- k_7 → hole pathway
- k_0 → hole + electron pathway

Discussion and Comparism

- Reaction rates and equilibrium constants differ from literature
- Reasons:
  - Literature used higher concentration
  - Much lower Rhodamine B dye as a surrogate for pharmaceutical and household wastewater
  - The constants are governed by many factors
- Correlated well with our data
- Reproducible

Conclusion

- Reaction rates and equilibrium constants differ from literature and are reproducible
- Reasons:
  - Literature used higher concentration
  - Much lower Rhodamine B dye as a surrogate for pharmaceutical and household wastewater
  - A simplified model was generated
  - Results represent a pseudo second order rate equation
  - Values for k1k4 were obtained

Recommendation

- Testing the effect of changing the pH on Cu2+ scavenging the electrons
- Testing the degradation rate of RB under doped photo catalysts
- Testing the degradation rate of other organic pollutants
- Testing the degradation rate vs wavelength of UV irradiation
- OH. radicals production by hole

Thanks For Listening

481424 Can Kriging Replace ODE-Based Kinetic Modeling ? a Hydrocracking Case Study

Abstract:

Can Kriging replace ODE-based kinetic modeling ? A hydrocracking case study

Valentin Dallerit[1], Benoit Celse[1], Victor Costa[1]*

1. IFP Energies Nouvelles, Rond-Point de l'Echangeur de Solaize, 69360 Solaize, France

**E-mail: benoit.celse@ifpen.fr*

HIGHLIGHTS

- Comparison between kinetic model and Kriging model.
- Good predictions in both cases.
- Kinetic model can enforce constraints and trends making them more robust
- Kriging models are more accurate but less robust. They are also built with uncertainties

INTRODUCTION

Catalytic hydrocracking is used to convert heavy Vacuum Gas Oil (VGO) residue to more valuable middle distillate and/or naphtha cuts. Hydrocracking of VGO residue is performed in a two-step process:

1) a hydrotreatment step in the first reactor (R1), which serves mainly to remove nitrogen and sulfur compounds from the feedstock (Becker et al. 2015);

2) a hydrocracking step in the second reactor (R2), which performs the main hydrocracking reactions on a zeolite catalyst (Per Julian Becker et al. 2016).

Models are developed to optimize process design and operating conditions in order to maximize desired cuts and product characteristics.

Historically, kinetic model (based on ODE) were used to predict the nitrogen content on the output of the first reactor. The aim of this paper is to present a new kind of model based of Kriging (also called Gaussian process regression (Kleijnen, Jack P. C. 2009)) and compare the results with the original kinetic model.

MATERIALS & METHODS

A kinetic model of the hydrodenitrogenation (HDN) is developed containing 11 parameters. The model parameters are fitted using 115 experimental runs in a pilot plant, with 15 feedstocks. The feedstocks have different geological origins (Arabian, Ural, Forcados...) or are produced in different conversion processes (coker, deasphalting...). Feedstock characteristics (SIMDIS, sulfur/nitrogen content, aromatic carbon content etc.) are known. The reactor temperature varies from 360 to 420°C, and the liquid hourly space velocity (LHSV) varies from 1.0 to 3.5 h^{-1}. Experiments were performed at 90, 110 and 140 bar with a hydrogen-to-oil ratio of 600 to 1400 l/l. An additional set of experimental runs was used to validate the model.

The same data are used in order to fit a model based on Kriging. Kriging is originally an exact interpolation method. However, some improvements to the method are used to allow the use of noisy data and avoid the model to pass exactly through the experimental data.

Kriging is mainly used on data using spatial coordinates (geophysics). The use of this method to predict physicochemical properties is a real innovation. This work shows the adaptability of kriging to nonlinear regressions.

Another advantage of Kriging is that it provides a measurement of predicted values uncertainties unlike other interpolation methods. Each predicted value is associated to a confident interval, the width of which depends on the related experimental conditions. The uncertainty is equal to the noise on data used to fit the model and increase with the distance to the calibration database. This distance, defined by the Kriging, give more weight to sensible inputs and can be used as a standalone distance to compute the distance between a new data and the calibration database. This can be used as a tool in order to detect extrapolation.

RESULTS & DISCUSSIONS

The comparison with the interpolation model for the experimental validation database on the nitrogen content was carried out. It shows that the interpolation model is more accurate but less robust than the kinetic model when the distance between the point and the calibration database increase. In this case, the kinetic model provides better simulation results.

However, the kriging model provides an uncertainty estimation. This kind of method is very easy to use and parameters tuning is very fast. It can be very useful for processes where the chemical behavior is not well known. It can also be used as a tool for design of experiments. New points are added where the uncertainty is maximal.

Bibliography

Becker, Per Julian; Celse, Benoit; Guillaume, Denis; Dulot, Hugues; Costa, Victor (2015) Hydrotreatment modeling for a variety of VGO feedstocks: A continuous lumping approach. In : Fuel, vol. 139, p. 133–143. DOI: 10.1016/j.fuel.2014.08.032.

Kleijnen, Jack P. C. (2009) Kriging metamodeling in simulation: A review. In : European Journal of Operational Research, vol. 192, n° 3, p. 707–716. DOI: 10.1016/j.ejor.2007.10.013.

Per Julian Becker; B. Celse; D. Guillaume; V. Costa; E. Guillon; G. Pirngruber (2016) A continuous lumping model for hydrocracking on a zeolite catalysts: model development and parameter identification. In : Fuel, vol. 164, p. 73–82.

478826 Facility Layout Optimization of LNG-FSRU System

Abstract:

Facility Layout Optimization of LNG-FSRU System

Chenxi Ji, Monir Ahammad and Sam Mannan

Mary Kay O'Connor Process Safety Center

Artie McFerrin Department of Chemical Engineering

Texas A&M University

College Station, Texas 77843-3122, USA

+1(281)684-9178, chenxiji@tamu.edu

ABSTRACT

This research presents a quantitative assessment of layout optimization of liquefied natural gas (LNG) floating storage and regasification unit (FSRU) system.

Compared to other loading/unloading facilities, the LNG-FSRU system has advantages on both building and operation phases such as cost-effectiveness and time efficiency, minimal impacts to the surrounding environment. Thus a quantitative study on the optimal facility sitting has been performed on the LNG-FSRU system involving simultaneously safe and economic objectives based on the Multi- Attribute Decision Analysis (MADA). Hazards are identified by Hazard and Operability (HAZOP) study, and then fault tree analysis and event tree analysis are adopted in order to monitor and control the safety performance of the LNG-FSRU system and to assess probabilities of the outcomes and overall system analysis, respectively. Three factors, i.e., operational safety index, transportation safety index and cost index are evaluated to establish utility function. In addition, sub-factors such as cryogenic risk, fire and explosion risk, ship collision risk and stranding risk were analyzed. The proposed approach was applied to a case study for the LNG-FSRU infrastructure layout parameters to optimize the construction location.

To conclude, R and AgenaRisk were applied to simulate the outputs of different scenarios so that the decision can be made by maximum utility value.

479970 Chemical Engineering's Role in the Upcoming Energy Transition

Abstract:

Concerns over climate change and lower cost Solar-PV are driving change Energy and Chemicals. This poster will present issues to be addressed in managing the Energy Transition to a more sustainable economy, and impact on the chemical engineering profession and its evolving technologies.

480438 Compressor Fouling Monitoring Methods Identify Fouling Mechanism Changes to Direct Mitigation Performance Improvement

Abstract:

Fouling within the olefins plant cracked gas compressor is a relatively common occurrence. Many gas crackers report some concern with compressor fouling. Significant percentages of liquid and mixed feed crackers also report fouling concerns. In some plants, the fouling occurs to a degree that imposes minimal operational limitations. The primary concern in these plants is an increase in energy consumption, and the fear of limitations if or when something causes the rate of fouling to increase. However, other plants consistently struggle with fouling that limits operational reliability and throughput capacity. A variety of fouling mitigation methods are available: internal coatings, specialty wash-oils and antifoulant additives. In most cases, the fouling is caused by a reactive polymerization mechanism. Quite often, this is a free radical mechanism, but Diels Alder mechanisms also occur. Changes in feedstock and/or changes in cracking severity can change or shift the predominance of the foulant mechanism. As the fouling mechanism changes, the mitigation method effectiveness may also change. In this paper, the fundamental monitoring methods to aid the evaluation of compressor performance and possible changes in fouling mechanism to direct changes in fouling mitigation techniques. Laboratory tests on process stream samples and operational data analysis procedures will be demonstrated. A case history will be presented to demonstrate how this methodology was used to facilitate performance improvement with an additive antifoulant mitigation method.

THE EVOLUTION OF LIGHT ENDS FRACTIONATION FOULING CONTROL: FROM MECHANISM TO MITIGATION, THREE DECADES OF DRAMATIC IMPROVEMENT

AIChE Poster Session, Paper #53am

M. David Smith
Industry Technical Consultant, Petrochemicals
Nalco Champion, an Ecolab Company

Introduction

The ethylene industry has seen many changes since the first steam cracker was designed and built in the 1940s, as olefins plants became larger and more complex and ethylene production worldwide grew; between the 1950s and present day, individual "world-scale" ethylene plants have increased in capacity by a factor of 20 and global production capacity has increased by nearly 200%[1,2]. In addition to capacity increases, plant operation has improved dramatically, especially over the past three decades. One study found that in the period from 1993 to 2009, the average run length for ethylene units between turnarounds increased by nearly 80% in North America and 45% worldwide[3].

There are various factors that impact the turnaround interval, but the most common factors are regulator inspection mandates; major maintenance, expansion, and equipment repair or replacement; and fouling that impacts operation[4]. This paper will focus on the innovations in fouling mitigation, specifically in the light ends fractionation area, that ethylene producers have employed to extend run lengths from as short as 1-2 years to as much as 8-10 years in the past three decades, while operating more safely, efficiently, and reliably. These innovations are comprised of adjustments made in three main areas: mechanical, operational, and chemical.

Light Ends Fractionation Fouling

The light ends fractionation section, sometimes called the cold ends, product separation, or recovery area, is a series of distillation towers that separate by boiling point the products produced in the cracking furnaces. The number and configuration of the various towers can vary from plant to plant, depending on the plant's vintage, cracking feedstock, licensor, and debottlenecking modifications; however, light ends sections typically involve a demethanizer (removing C_1s and lighter as the overhead stream), deethanizer (C_2s), depropanizer (C_3s), and debutanizer (C_4s). Variations typically are based on which tower is the first in the distillation train; for example, the "conventional" design had the demethanizer first. Other configurations include a deethanizer-first in gas crackers and depropanizer-first in liquid crackers.

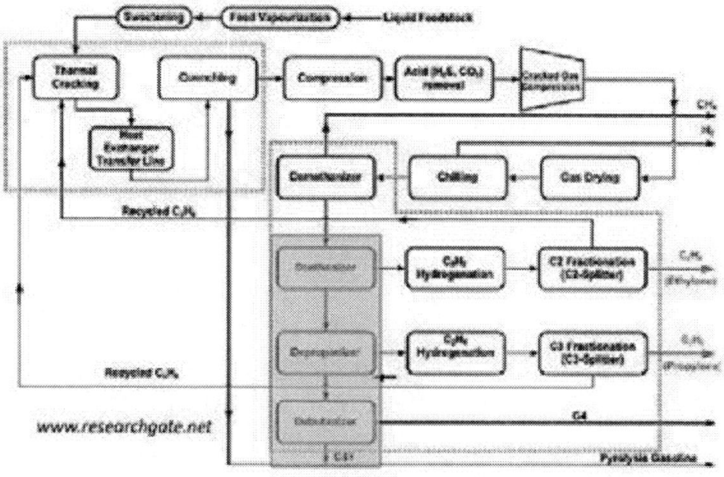

Figure 1: Typical ethylene unit diagram, with a "conventional" configuration

Fouling in the light ends occurs in the towers, reboilers, and associated piping, and is the result of free radical polymerization of reactive monomers such as 1,3-butadiene and isoprene, with occasional Diels-Alder reactions also occurring. Primarily, however, the foulant material consists of polybutadiene[5]. Free radicals are initiated by heat or by peroxy radicals that are themselves formed by either heat or a metal catalyst. As the radical reacts with a monomer (propagation), the polymer chain grows; when the chain becomes long enough or a change in solvency occurs, the material can lay down on piping or equipment[6].

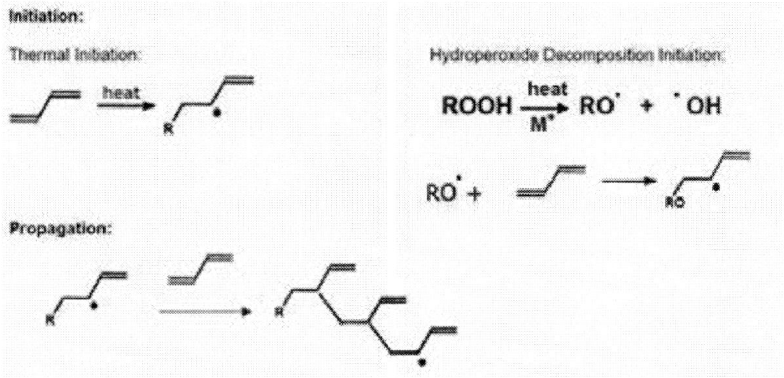

Figure 2: Free radical polymerization – initiation and propagation steps

The severity of the fouling is dependent on several factors: concentration of the reactive monomer, temperature, residence time, solvency, contaminates, pH, and oxygen[7]. These factors can vary based on furnace feedstock, production rates, equipment design, recycle streams from other units, metallurgy, and other influences. The deleterious impacts of fouling in the light ends area include decreased heat transfer, reduced equipment reliability, increased cleaning costs, off-spec product, decreased throughput, and, ultimately plant shutdown. In the case of the formation of aggressive, proliferous polymerization – pyrophoric "popcorn" polymer – the result can be bent or even breached metal and loss of containment[8].

The significance of these impacts provides ample incentive for ethylene producers to minimize the impacts of fouling in the light ends area. To accomplish this, they have utilized mechanical, operational, and chemical innovations.

Mechanical and Operational Innovations

Over the years, plant design and other mechanical innovations have been implemented to extend unit run length and improve reliability. One such implement was to install spare equipment: spare reboilers can be swapped without disrupting production when one reboiler fouls; the fouled reboiler can be cleaned online and the process repeated. In some cases, entire distillation towers or sections of towers were spared as well to facilitate extended runs.

Light ends reboilers vary in design; among other differences, they can be horizontal or vertical, and the process can be on the shell side or the tube side. Thorough cleaning of the offline reboiler is more easily done in-situ in a vertical reboiler with process on the tube side; in this case, the tubes can be hydroblasted efficiently, provided a sufficiently high-pressure water jet is used. Reboilers with the process on the shell side are more difficult to hydroblast; in these exchangers, the tube bundle typically must be sent offsite to have the polymer burned off the tubes ("baking"). Thorough polymer removal not only helps extend the run length of the reboiler, it lowers the chance for leftover polymer to be carried into the tower and foul trays or downcomers.

As mentioned above, peroxide decomposition can be catalyzed by metals, and smooth surfaces accumulate polymer less readily than abraded surfaces. To reduce fouling, some ethylene producers have replaced carbon steel column internals with stainless steel. In addition to metallurgy conversion, there are other tray design changes that have reduced fouling in light ends towers[9].

Temperature is a significant contributor to fouling severity: higher temperatures greatly increase polymerization reaction rate, and there is a noted correlation between temperature and reboiler run length[10]. In older vintage depropanizers, temperatures ran upwards of 200°F; in newer plants, temperatures have been reduced to a maximum of approximately 180°F using various design innovations such as desuperheated steam, changing heating medium, dual-pressure columns, and varied distillation train configuration, among others[11].

Oxygen present in the system can form peroxides and promote popcorn polymer[12]. While porous material such as Teflon™ is compatible with butadiene and therefore has been used historically in gaskets, it can hold oxygen or absorb reactive monomer and therefore should not be used in butadiene service[13]. Replacing the Teflon™ gaskets with a non-porous substitute has helped extend light ends tower and reboiler run length and prevented popcorn formation.

An often-overlooked strain on reboiler performance is the steam condensate return system. As plants debottleneck and increase production capacity, often they increase the amount of condensate going back to the boiler but do not re-size the piping infrastructure; thus, the return header pressure increases. In some cases, the backpressure on the condensate causes it to "back up" into the reboiler. This decreases the available heat transfer area, working the reboiler harder and potentially decreasing the run length due to actual or perceived fouling. Another cause of condensate backup can be ineffective "pumping traps" in the condensate removal system. Some producers have removed this strain by installing condensate pumps, decreasing the fouling severity and improving reboiler performance.

Procedures for isolating equipment when off-line have been developed to help inhibit fouling formation. Nitrogen-blanketing equipment immediately following cleaning prevents rust formation that can catalyze the formation of free radicals.

A final operational change that has helped improve reliability is improved inspection and analytical methods. Tower gamma scans can pinpoint where in a tower fouling or mechanical damage is present[14]. Analytical methods can determine the composition of a foulant sample, helping elucidate where the foulant was formed and if the material is popcorn or normal polymer.

This list is not exhaustive, but it demonstrates the extensive number of adjustments and modifications ethylene producers, licensors, and other support industries have implemented to help extend the turnaround interval, improve operational reliability and efficiency, and reduce costs.

Chemical Innovations

In addition to mechanical and operational upgrades, ethylene producers have used chemical treatment breakthroughs to further reduce the impacts of fouling in the light ends area; the use of antifoulants has become standard in the industry[15], and it is estimated that over 95% of all light ends distillation trains use some form of chemical antifoulant additive. The commonly used chemical treatment additives in light ends include the following[16,17]:

- Antioxidants
    - Primary
    - Secondary
    - Oxygen scavengers
- Antipolymerants
- Dispersants
- Passivators
- Stabilizers

Primary antioxidants scavenge peroxy radicals. They essentially donate a proton to "cap" the peroxy radical. They can also react with carbon-centered free radicals, but the reaction is very slow. Secondary antioxidants decompose peroxide radicals via a mechanism that does not produce free radicals. Oxygen scavengers react with oxygen and remove it from the system. The first antifoulants widely used in light ends systems were mainly primary antioxidants such as phenylenediamines (PDAs) and diethylhydroxylamine (DEHA).

Antipolymerants are stable free radicals (SFRs). They react very quickly with carbon-centered free radicals to terminate the polymerization reaction. Because they react so quickly, very low dosages are typically required. Patented in the late 1980s for use in light ends[18], SFR inhibitors greatly reduced fouling and significantly increased equipment run lengths[19].

Dispersants do not impede the free radical polymerization reaction, but rather prevent polymeric particulates from agglomerating and depositing on equipment. In some cases, they can effect a reverse deposition by breaking up agglomerates or loosening existing deposits, resulting in a recovery of performance; this is dependent on the tenacious nature of the foulant.

Passivators change the oxidation state of iron so that it forms a metal protective surface with a lower catalyst activity to inhibit the formation of peroxy radicals. The result is better start-of-run performance and reduced fouling in-run. Passivation with an inhibitor can deactivate popcorn "seeds" and reduce the risk of popcorn formation.

Stabilizers are typically primary antioxidants used to inhibit the formation of explosive polyperoxides and/or gums or polymer in product streams. Sometimes their use is mandated: United States Department of Transportation regulations specify that crude C_4 streams (debutanizer overhead product) must be "inhibited" or "stabilized"[20]. Also, storage tank and pipeline fouling can result in expensive cleaning costs and even plant shutdown. Use of a stabilizer can therefore meet regulatory approval and increase equipment and pipeline reliability.

The use of these chemical treatment additives has helped make dramatic improvements in the mitigation of polymerization in the light ends area.

The Path Ahead

The impact of shale gas in North America on ethylene production has been called a "revolution", reigniting petrochemical investment with at least 7 new ethylene plants under construction along the U.S. Gulf Coast and others awaiting final investment decision (FID)[21]; many other facilities have expanded capacity in existing units. The impact is spreading worldwide as U.S. shale gas feedstocks are now being exported around the globe[22]. The impacts on light ends fouling have been well documented in the literature, including in several Ethylene Producers Conferences: plants designed for cracking liquids have moved to light feedstocks that are plentiful due to shale gas. This shift, along with upstream treatment additives remaining in the gas feed, has increased the fouling severity in the light ends fractionation area[22].

Even as fouling severities have increased and plants are moving into unprecedented operating regimes, ethylene producers are still trying to continue to increase throughput, reduce energy consumption, improve reliability, and extend turnaround intervals – all while operating more safely and environmentally friendly. To meet these growing demands, continued innovation will be required in each of the areas discussed above.

Summary

Over the past 30 years, the petrochemical industry has experienced considerable growth in demand, and supply has been expanded to meet the still-growing petrochemical marketplace. It is a marketplace driven by mechanical, operational, and chemical treatment innovation that has resulted in remarkable improvements in safety, reliability, environmental stewardship, production, and on-line operation.

It truly has been Three Decades of Dramatic Improvement. With a revitalized marketplace, the next three decades should see even more growth and innovation.

Acknowledgements

The author would like to thank Ted Arnst, Debby Rossana, Javier Florencio, and Jessica Hancock for valued assistance in the preparation of this paper.

References

1. GlobalData (2016), "Global Ethylene Capacity and Capital Expenditure Outlook – US and China to Lead Ethylene Industry Expansion", www.globaldata.com.
2. Kapur, Sanjeev, *Handbook of Petrochemicals Production Processes (2005)*, Robert A. Meyers, ed. New York, NY, McGraw-Hill, pp. 6.3-6.5.
3. Cagnolatti, Claire L. and Richard B. Jones (2012), "Analysis of Turnarounds in Olefin Plants," AIChE Spring National Meeting, Ethylene Producers Conference, Paper 127a.
4. de Braun, C.J. (1994), "Extended Operating Cycles in Ethylene Plants,", AIChE Spring National Meeting, Ethylene Producers Conference, Article 23a.
5. Hancock, J.M., A.W. Van Zijl, I. Robson, J. Vachon, and A. Hashmi, "A Chemist's Perspective on Organic Fouling in Ethylene Operations," *Hydrocarbon Processing*, June 2014, p. 64.
6. Snider, Sheri (2006), "Ethylene Plant Cracked Gas Compressor Fouling", AIChE Spring National Meeting, Ethylene Producers Conference, Session 194.
7. Vachon, J., J.M. Hancock, and I. Robson, "A Chemist's Perspective on Organic Fouling in Ethylene Operations: Update," *Hydrocarbon Processing*, October 2015, p.49.
8. Rossana, Debby and Jessica M. Hancock (2014), "Popcorn Polymer Formation in Light Ends Units," AIChE Spring National Meeting, Ethylene Producers Conference, Session 36.
9. Kolmetz, Karl, Andrew W. Sloley, Timothy M. Zygula, Wai Kiong Ng, and Peter W. Faessler (2004), "Design Guidelines for Distillation Columns in Fouling Service," AIChE Spring National Meeting, Ethylene Producers Conference, Session 66.
10. Vachon, J. et.al., p. 50.
11. De Haan, Stephen, Barbara Stancato, and Brian Sullivan (2009), "The Impact of Fouling and Corrosion on Ethylene Plant Efficiency," AIChE Spring National Meeting, Ethylene Producers Conference, Session 110.
12. Nowowiejski, George and John A. Reid (2003), "An Overview of Oxygenates in Olefins Units in Relation to Corrosion, Fouling, Product Specifications, and Safety," AIChE Spring National Meeting, Ethylene Producers Conference, Session 56.
13. Hancock, Jessica M., Debby Rossana, Stefanus Korf, "Popcorn in Light Ends," *Hydrocarbon Engineering*, December 2014.
14. Zister, Henry Z. and Chuck Winfield, "Use Downcomer Gamma Scans to Troubleshoot Your Process," *Chemical Engineering Progress*, January 2017, pp. 28-38.
15. Kim, Grace and Morgan Stante (2014), "Inventory and Fate of Chemical Additives to Improve Plant Reliability," AIChE Spring National Meeting, Ethylene Producers Conference, Paper 36e.
16. Ibid.
17. Rossana, Debby and Jessica M. Hancock.
18. Ferrell, Thomas M, "Method for Controlling Fouling of Hydrocarbon Compositions Containing Olefinic Compounds," US Patent 4670131 A.
19. Hancock, J.M. et.al., p.65.
20. United States Federal Register, 46 CFR 151, Part 151.

21. Zinger, Steve, Vikki Medley, and Anne Keller (2015), "The New World Order in the Olefins Industry," AIChE Spring National Meeting, Ethylene Producers Conference, Session 47.
22. Lippe, Dan, "US Midstream Industry Anticipates Surge in Ethane Demand, Exports," *Oil & Gas Journal*, November 2016.
23. Lordo, Samuel A. and M. David Smith, "Oil Field Additives and Ethylene Feedstock Contaminants," AIChE Spring National Meeting, Ethylene Producers Conference, Paper 126b.

480911 Thermodynamic Parameter Regression Using Open Source Optimizers

Abstract:

Parameterization of activity models and equations of state for multicomponent systems is a complicated endeavor when the number of system components is high. Even when restricting interaction parameters to the binary level, the inherent non-linearity of thermodynamic models often leads to local solutions that are not globally optimal. In this talk, we examine the quality of open source optimizers when applied to thermodynamic parameter regression tasks, using the Python programming environment as a framework to drive Aspen Properties calculations. In particular, we consider techniques designed for discovering global optima (e.g., Particle Swarm Optimization). Finally, we provide practical examples relevant to Acid Gas Treating, such as multicomponent aqueous amine solvent systems.

481382 Latest Technology of Double Disc Gate Valves for Main Transfer Line and Decoke Line in the Ethylene Service

Abstract:

The company ZMK Technologies GmbH from Germany would like to present a paper at your ethylene producers' conference in San Antonio.

The purpose of the paper is to show the audience the latest technology and further development of the well known double disc type main transfer line and decoke line valves for steam cracker/ethylene producing plants.

All developments basically ensure that the maintenance as well as downtime of the special valves for main transfer line and decoke line application can be reduced quite drastically.

Attached please find a brief indication of what has been improved during the last two years on such special equipment.

To fulfil you general expectations we are currently in touch with ExxonMobil, BP as well as LyondellBasell to involve a manufacturing company as co-author.

The company ZMK Technologies GmbH from Germany would like to present a paper at your ethylene producers' conference in San Antonio.
The purpose of the paper is to show the audience the latest technology and further development of the well known double disc type main transfer line and decoke line valves for steam cracker/ethylene producing plants.
All developments basically ensure that the maintenance as well as downtime of the special valves for main transfer line and decoke line application can be reduced quite drastically.
Attached please find a brief indication of what has been improved during the last two years on such special equipment.
To fulfil you general expectations we are currently in touch with ExxonMobil, BP as well as LyondellBasell to involve a manufacturing company as co-author.

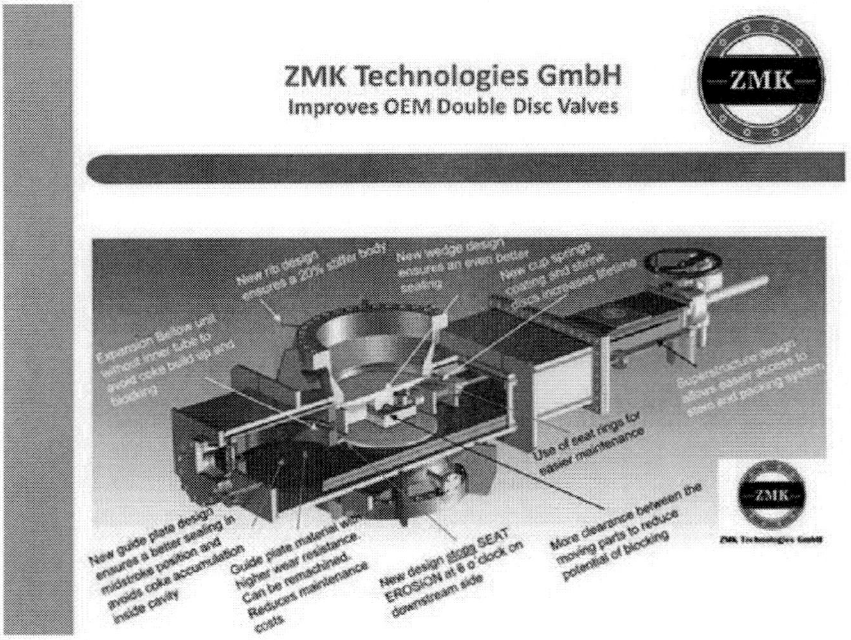

478215 Comprehensive Review on Hazards Related to Lithium-Ion Batteries Technology

Abstract:

With the recent developments on renewable energies, the capacity for energy storage represents a growing area for improvement. Rechargeable batteries have been recently improved in terms of durability, energy density, size and volume. Lithium-ion batteries represent one of the best examples of improvement of energy storage technologies; these have been widely used in a diverse field of applications, from common electronic devices used by millions of people around the world to electric and hybrid vehicles as alternative for conventional fuel-based transportation. Research on the improvement of the design of Lithium-ion batteries is constantly on-going and new materials are being investigated in order to improve the quality of the batteries in terms of efficiency and charge cycles. However, Lithium-ion batteries represent considerable hazards to customers as demonstrated by a series of significant incidents that have occurred in different battery cells involving overheating, fires and explosions in different devices.

Due to the exposure of these batteries, risks related to thermal runaway phenomena are presented in order to propose different approaches to follow in order to investigate causes, prevention measures and design changes to prevent these types of incidents. Thus, a comprehensive presentation of the current status of the safety research on lithium/ion batteries thermal runaway is presented. An overview of the possible promoters for the thermal event is presented by studying the kinetic characteristics of the phenomena as well as the different possible approaches to study this phenomenon such as molecular simulations to understand mechanisms and proposes related decomposition paths.

478532 Theoretical Evaluation of the Reactivity for Hydroxylamine System using a Kinetic Model Approach

Abstract:

Abstract: From 1980 to 2001 the U.S Chemical Safety and Hazard Investigation Board (CSB) reported 167 incidents involving reactive chemicals occurred in the U.S. According to the report 35% of these incidents where caused by thermal runaway reactions. The Hydroxylamine (HA) compound has been involved in two of these tragic accidents. Identification and understanding runaway reactions hazards help in preventing the recurrence of similar events. It is also essential to acquire knowledge of detailed chemical kinetics of energetic materials in order to optimize current processes.

Understanding the mechanisms of how a runaway reaction propagates help to identify conditions needed to control or mitigate these reactions. The aim of this research is to develop a robust computational methodology using the reactivity of the hydroxylamine as example to prevent the incident recurrence related to runaway reactions. Detailed kinetic model, *i.e.,*a series of elementary reactions for HA decomposition were developed. This model can be applied to ensure reliable scale-up of theoretical data to industrial simulations.

This study proposes a methodology consisting of three steps; first, computational predictions of the reaction mechanisms and calculation of thermodynamic properties of the reactions involved in the HA decomposition using GAUSSIAN. Second, the estimation of individual reaction rate constants. Variational Transition State Theory (VTST) using software as POLYRATE and GAUSSIAN is obtained from the literature where the application in this step consist in calculate the transition state and the activation energy where the reactions have occurred. Finally to study industrial scale operations, CFD simulations using ANSYS-FLUENT is performed in order to evaluate the behavior of the HA for an industrial process. From the obtained results, it is possible to estimate the accuracy of the proposed methodology to identify and quantify runaway reaction hazards.

Keywords: Runaway reactions, hydroxylamine, activation energy and industrial simulations.

Abstract for 2017 AIChE Spring Meeting and 13th Global Congress on Process Safety March 26-30, 2017 San Antonio, TX

Application of the Density Functional Theory (DFT) to Study the Initiation Decomposition of Nitroaromatic Explosive: 1,3-Diamino-2,4,6-Trinitrobenzene (DATB)

Ingry Ruiz[1], Lisa Perez[2], Bin Zhang[1] and M. Sam Mannan[1]

[1]Mary Kay O'Connor Process Safety Center, Artie McFerrin Department of Chemical Engineering Texas A&M University, College Station, Texas 77843-3122, USA

[2] Laboratory for Molecular Simulation, Texas A&M University, College Station, Texas 77843, USA

+1(979) 341-9220, ingryruiz27@gmail.com

Abstract: Nitroaromatic compounds are organic molecules that have at least one nitro group (-NO_2) attached to an aromatic ring. These compounds have long played an important role as explosives, rocket propellants, pesticides, and fuels. Hence, understand and analyze the decomposition behaviors of nitroaromatic explosives are very important in order to control the hazards of these materials, use them safely and also it may provide novel ways to design new materials. The decomposition of nitroaromatic explosives consists of three main stages: initiation, combustion and detonation. The last two steps are fairly well understood and it is clear that the initiation of decomposition is a complex process. In fact, the study on initiation is a significant challenge when several decomposition mechanisms are possible and many reactions products are produced. Thus, the development of tools to analyze the thermal and kinetic aspects of gas phase decomposition mechanisms have been promoted and which these existing tools are mostly based on empirical rules.

However, the decomposition mechanisms are closely linked to the explosive properties of energetic materials, for that reason, any approach that gives access to detailed mechanisms of their decomposition could contribute significant improvement in chemical hazard prevention. In this way theoretical chemistry opens new perspective. Recently, the study of decomposition mechanisms and properties of the energetic materials are based on empirical considerations due to experimental data are not always available. Many researches on the unimolecular decomposition of some nitroaromatic explosive have been undertaken (*e.g.* Picric Acid, TNT, MATB, and TATB) using computational tools based on a quantum-mechanical description.

In the nitroaromatic group we find the 1,3-Diamino-2,4,6-trinitrobenzene (DATB), $C_6H_5N_5O_6$, which production began in 1960. This compound is considered a thermally stable explosive and is used in modern ordinance, space and nuclear fields. However, investigations regarding the decomposition mechanisms have been a considerably slow progress. Therefore, the study of the initiation process of DATB is a challenge due to limited information related to initial decomposition mechanisms of detonation because through experimental methods it is difficult to distinguish between competing chemical reactions during the initial steps of the rapid decomposition of DATB. For this reason, the objective of this research is to study the initiation decomposition of DATB, and provide first-principles fundamental of the initial stage of the detonation, through the calculation of the thermodynamic and kinetics parameters, and mechanisms of its thermal decomposition using a theoretical method.

Here, we have postulated four possible pathways for DATB using the most common channels for Nitroarenes. These are: C-NO_2 Homolysis, NO Elimination, HONO Elimination and H2O Elimination. In these, the aromatic ring is kept intact due to results from previous works showing that the aromatic ring fission does not take place until most of the attached substituents are removed. It is believed that the ring fission is a process that most probably takes place during detonation. The results of DATB will compared with the behavior of other nitroaromatic such as TNT and TATB by default, all the interpretation will most often be based on speculations.

All the calculations will be carried out using density functional theory (DFT) which is based on the Hohenberg-Kohn theorem as implemented in the Gaussian 09 software. The Becke three-parameter hybrid and Lee, Yang, and Parr correlation (B3LYP) at the 6-311++G* basis set. Also, computational temperature analysis was developed to estimate the temperature dependence of all possible pathways during DATB decomposition. With this analysis, we want estimate the most favorable pathway that leads to DATB detonation and the temperature range during which each pathway becomes dominate.

Keywords: DFT, Nitroaromatic explosive, Decomposition mechanisms, DATB, Computational Quantum Chemistry,

479769 Bioelectrochemical Detection of Endocrine Disrupting Compounds

Abstract:

Endocrine disrupting compounds are ubiquitous, originating from pesticides, plasticizers, and pharmaceuticals, among other sources. Though the full impact of these compounds has not been fully elucidated, they are implicated in diseases such as obesity, diabetes, and cancer. The list of chemicals that disrupt normal hormone function is growing at an alarming rate, making it crucial to identify sources of contamination and new compounds that display this ability. We have developed new detection strategy for endocrine disrupting compounds that is both fast and portable. This method requires no specialized skills to perform based on a native estrogen receptor construct. Rather than responding to individual components, this system reports the total estrogenic activity of a sample using the most relevant biological receptor. This approach should be broadly applicable to the detection of chemically dissimilar classes of compounds that bind to a single receptor.

479982

Production of Hydrogen by Methane Decomposition Using Fe-supported Alumina Catalysts

Megumu Inaba, *Zhanguo Zhang*, *Koichi Matsuoka*

National Institute of Advanced Industrial Science and Technology (AIST),
Tsukuba, Ibaraki, JAPAN
Megumu Inaba: mg.inaba@aist.go.jp
Zhanguo Zhang: z.zhang@aist.go.jp
Koichi Matsuoka: koichi-matsuoka@aist.go.jp

Keywords

Catalysts, Hydrogen Production & Storage and Reaction Mechanism

Abstract

In reaction of methane decomposition using Fe-supported alumina catalyst, higher reaction temperature led to shorter induction period before methane conversion starts and slightly higher maximum in methane conversion, but led to larger decrease in methane conversion. Moreover, higher flow rate of methane did not affect the maximum in methane conversion rate so significantly, suggesting that this reaction reached equilibrium state. Higher flow rate led to larger decrease in methane conversion, shorter catalytic lifetime and less amount of deposited carbon. The presence of carbon dioxide in methane gas inhibited reduction of iron oxide, and as a result, longer induction period and higher reaction temperature are needed for methane conversion.

Introduction

Methane is the simplest hydrocarbon, and mainly, can be obtained from natural gas, petroleum and methane hydrate. However, methane is contained in biogas which can be obtained by fermentation of biomass such as excrement and sewage sludge. In this case, carbon dioxide is contained with methane in biogas. Much attention has been paid to useful utilization of biomass, since biomass is considered to be renewable and carbon neutral. Global warming potential of methane is ca. 25 times higher than carbon dioxide. Emission of methane must be avoided to prevent global warming. Combustion is one of utilization methods of methane, but even by combustion, carbon dioxide, which also causes greenhouse effect, is emitted. Therefore, other method of methane utilization without emitting carbon dioxide is expected.

Hydrogen is expected to be clean energy source because combustion of hydrogen leads to formation of only water, and no carbon dioxide. As a method for production of hydrogen, steam-reforming of hydrocarbon has been known. However, this method leads to formation of carbon dioxide, and development of the method for production of hydrogen without formation of carbon dioxide is expected. As other method, decomposition of methane over catalysts has been studied quite extensively. It is well known that Ni catalysts are effective for this reaction [1]. However, Ni is toxic and somewhat expensive metal. On the other hand, Fe is nontoxic and cheap metal. Therefore, there will be many merits to use Fe as catalyst [2, 3]. For example, deposited carbon containing Fe after reaction can be disposed safely and easily. In both cases of Ni and Fe, reduced metals have catalytic activity for methane decomposition.

We tried production of hydrogen and carbon by methane decomposition using Fe-supported alumina catalyst, and investigated the effects of flow rate of methane and reaction temperature on catalytic performance during methane decomposition reaction. Moreover, effect of carbon dioxide with methane on catalytic performance was investigated, since biogas which can be obtained by fermentation of biomass contains not only methane but also carbon dioxide.

Experimental

Fe-supported alumina catalyst was obtained from The Japan Steel Works, LTD. (prepared by Süd-Chemie Catalysts Japan, Inc.). As characterization of catalyst before reaction, powder X-ray diffraction (XRD) analysis and X-ray Fluorescence (XRF) analysis were carried out. XRD analysis (Rigaku Corporation, MiniFlex600) was carried out to investigate the crystal structure of catalysts. The result of XRD analysis suggested that hematite (Fe_2O_3) and θ-alumina (Al_2O_3) coexist in Fe-supported alumina catalysts. XRF analysis (Rigaku Corporation, EDXL300) was also performed to determine the ratio of each element. From the result, content of Fe_2O_3 was calculated to ca. 44 wt. %.

Catalytic activity was measured with a fixed-bed reactor. Quartz reactor with inner diameter of 36 mm and length of 460 mm was used in lying condition. Catalyst weight was 0.2 g. Catalyst was mounted in quartz boat (25 mm width × 50 mm length × 12 mm depth) and the boat was set beside thermocouple in the quartz reactor. Reactor was heated to reaction temperature with heating rate of 10 °C/min in image furnace in flow of methane from start of heating. Methane has activity for reduction of iron oxide. As methane, pure methane and 1-3% CO_2-containing methane were used. Flow rate of methane was varied from 10 to 80 ml/min, and reaction temperature was varied from 670 to 760 °C. They were constant during reaction. Between quartz reactor and gas chromatography (GC), glass tube willed with silica gel particles was connected in order to adsorb water formed during reduction of iron oxides. Effluent gas was detected by GC (Shimadzu

Corporation, Gas Chromatograph GC-8A) equipped with thermal conductivity detector (TCD) and packed column (GL Sciences Inc., Molecular Sieve 5A). Argon was used as carrier gas. Temperature of GC column was set to 50 °C.

Under these conditions, change of rate of methane conversion with time-on-stream was observed. Mechanism of hydrogen production by decomposition of pure methane is considered to progress by following equation: $CH_4 \rightarrow C + 2H_2$. In the presence of CO_2, this reaction can be considered to progress as follows: $CH_4 + CO_2 \rightarrow 2H_2 + 2CO$. Methane conversion (%) was defined as follows: Methane conversion = (hydrogen concentration) / (200 − (hydrogen concentration)) × 100. Hydrogen concentration (%) was calculated from GC peak area. Basically, reaction time was fixed to 6 h.

Weight of catalyst before and after reaction was measured, and from these results, amount of deposited carbon was calculated. In this case, reaction was carried out until catalyst was almost deactivated.

Results and Discussion

Primarily, time dependence on methane conversion was investigated in condition that flow rate of methane was 20 ml/min and only pure methane was supplied from start of heating (Fig. 1). At 680 °C, methane was scarcely converted even after 6 h from start of reaction. At 690 °C, methane conversion was increased to 56.0 % after induction period of ca. 110 min. Induction period is time necessary for reduction of Fe_2O_3 by methane. Above 700 °C, induction period became shorter with increasing reaction temperature. Maximum values of methane conversion at 690, 700, 720 and 740 °C were 56.0, 59.9, 66.2 and 69.0 %, respectively. Difference in methane conversion by reaction temperature was not so significant, suggesting that this reaction may reach equilibrium state. Equilibrium value is varied by temperature: the higher temperature leads to higher equilibrium value. It seems that reaction temperature of 680 °C may not be sufficient temperature for reduction of Fe_2O_3 by methane with flow rate of 20 ml/min, since it is suggested that reduced FeO species have catalytic activity for methane decomposition. On the other hand, in all cases, methane conversion was gradually decreased after methane conversion reached maximum. The decrease from the maximum value to the value at 6 h after start of reaction at 690, 700, 720 and 740 °C were 3.2, 7.1, 11.5 and 17.8 %, respectively. These results suggested that higher reaction temperature leads to larger decrease in methane conversion with time-on-stream.

Next, time dependence on methane conversion was investigated in condition that flow rate of methane was 40 ml/min and pure methane was supplied from start of heating (Fig. 2). In this case, methane conversion could not proceed at 670 °C, while at 680 °C, methane conversion was increased to 51.5 % after induction period of ca. 40 min. Maximum values of methane conversion at 680, 690, 700, 720 and 740 °C were 51.5, 54.1, 56.7, 63.7 and 67.4 %, respectively. The maximum value at each temperature was similar with maximum at corresponding temperature, respectively (i.e., quantity of methane conversion was twice as much as at flow rate of 20 ml/min). Generally, induction period at methane flow rate of 40 ml/min was shorter than at 20 ml/min. In this case, also, after reaching the maximum of methane conversion, methane conversion was decreased with time-on-stream. The decrease from the maximum value to the value at 6 h after start of reaction at 680, 690, 700, 720 and 740 °C were 14.2, 21.9, 38.6, 52.4 and 58.3 %, respectively. The decrease was larger for methane flow rate of 40 ml/min than for 20 ml/min, and was larger for higher reaction temperature than for lower temperature. More amount of methane flow and higher reaction temperature led to the larger decrease in methane conversion with time-on-stream.

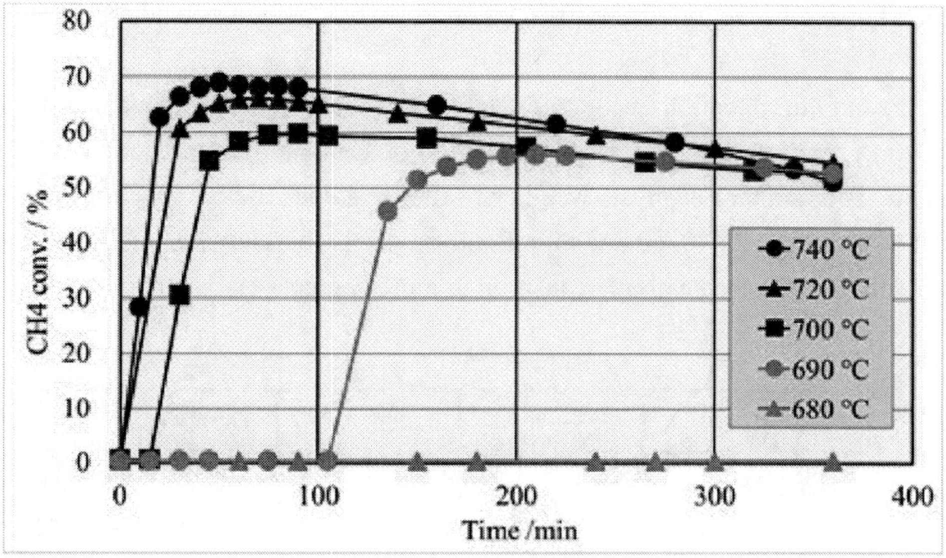

Fig.1 Change of methane conversion with time-on-stream.
Flow rate of pure methane = 20 ml/min.

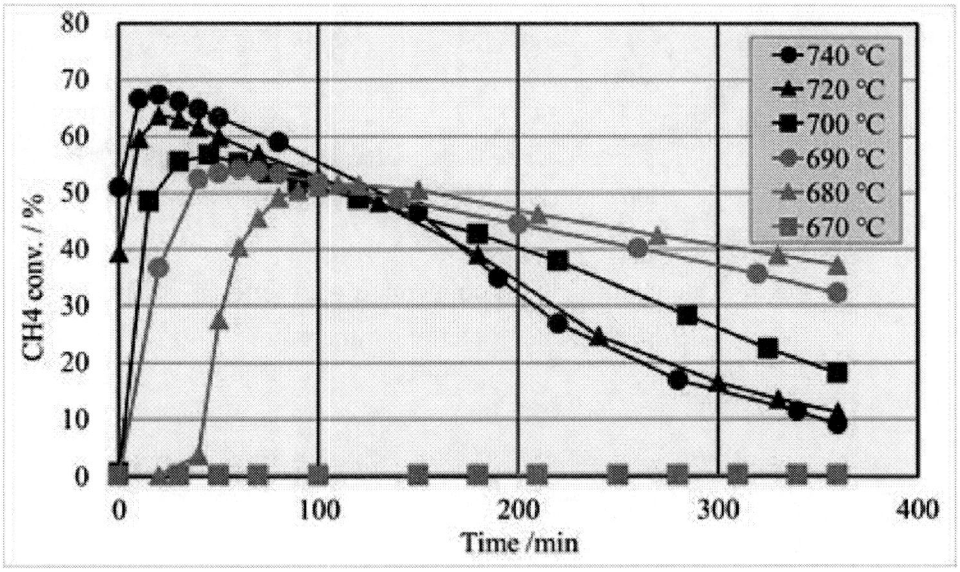

Fig. 2 Change of methane conversion with time-on-stream.
Flow rate of pure methane = 40 ml/min.

Effect of flow rate of methane on catalytic lifetime was investigated at 740 °C (Fig. 3). Flow rate was 10, 20, 40, 60 or 80 ml/min and pure methane was supplied from start of heating. In this case, catalytic reaction was carried out until catalysts were almost deactivated. In the case of flow rate of 10 ml/min, catalyst was deactivated after ca. 2000 min from start of reaction. Catalysts were deactivated after ca. 1000 and 500 min for methane flow rate of 20 and 40 ml/min, respectively. Order of catalytic lifetime was as follows: 10 > 20 > 40 > 60 > 80 ml/min, suggesting that faster flow rate led to shorter catalytic lifetime. In the cases of flow rate of 40, 60 and 80 ml/min, maximum value of methane conversion rate (58.7, 59.1 and 57.4, respectively) was lower than the cases of 10 and 20 ml/min (72.3 and 69.2, respectively), suggesting that, in these cases, methane decomposition reaction did not reach equilibrium state, since methane flow rate was too fast.

Before and after reaction, weight of catalyst was measured to investigate change in weight of catalyst during reaction (Fig.4). Carbon deposition on catalyst during reaction led to increase in catalyst weight. The ratios of catalysts weight of (after reaction)/(before reaction) were 20.6, 16.7, 13.0, 9.9 and 10.7 for 10, 20, 40, 60 and 80 ml/min, respectively. These results suggested that lower flow rate of methane enhanced carbon deposition in more degree.

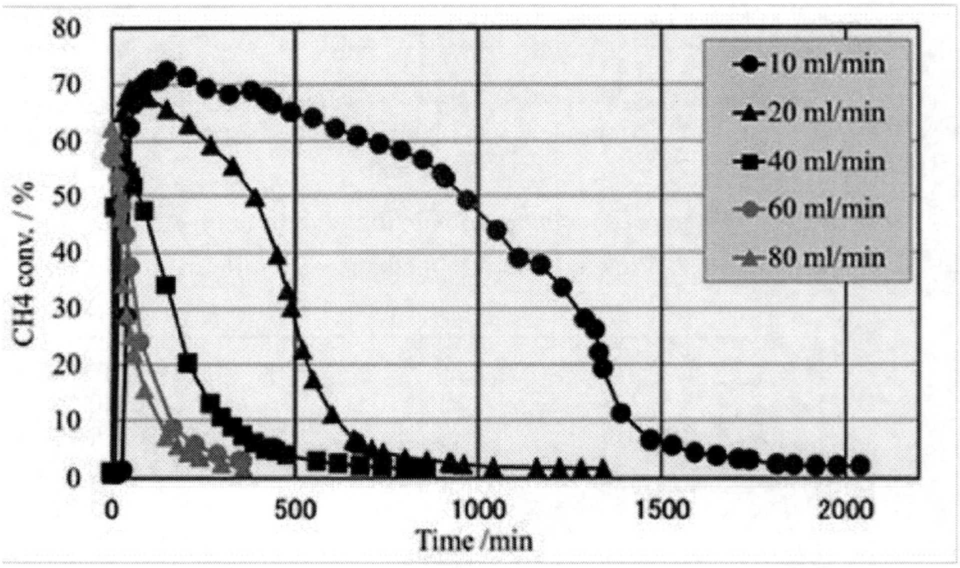

Fig. 3 Change of methane conversion with time-on-stream.
Pure methane, Reaction temperature = 740 °C.

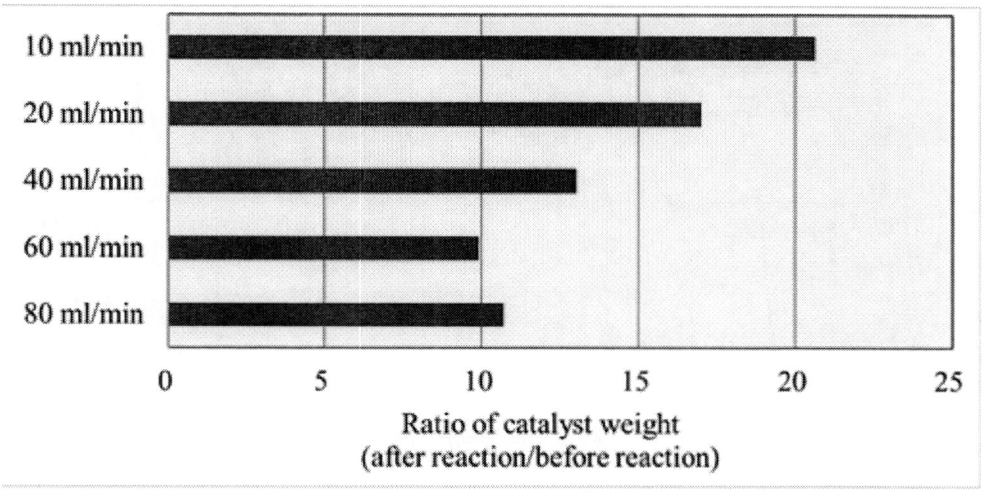

Fig. 4 Change in catalyst weight after reaction to deactivation.
Pure methane, Reaction temperature = 740 °C.

Effect of existence of CO_2 mixed with methane was investigated, since biogas obtained by fermentation of biomass contains CO_2 as well as methane. Time dependence on methane conversion was investigated in condition that flow rate of methane was 20 ml/min and CO_2-containing methane was supplied from start of heating (Fig. 5). To make a comparison, results of reaction of pure methane were also shown in Fig. 5. In the case of 1 % of CO_2 contained in methane, at 710 °C, methane was hardly converted even after 6 h from start of reaction. At 720 °C, methane conversion was increased to 60.5 % after induction period of ca. 110 min. Above 720 °C, higher reaction temperature led to shorter induction period and higher methane conversion. In the case of 2 % of CO_2 contained in methane, methane conversion scarcely occurred at 720 °C, while methane conversion was increased to 60.5 % after induction period of ca. 150 min at 740 °C. At 750 °C, methane conversion was increased to 68.3 % after induction period of ca. 30 min. In the case of 2%, also, higher reaction temperature led to shorter induction period and higher methane conversion. In the case of 3 % of CO_2, methane conversion scarcely occurred at even 750 °C, while methane conversion was increased to 71.5 % after induction period of ca. 60 min at 760 °C. In higher concentration of CO_2, higher reaction temperature was necessary for methane decomposition reaction. From these results, it is suggested that higher concentration of CO_2 inhibits the reduction of FeO species, and as a results, it needs higher temperature to reduce FeO species to active Fe species. However, at 740 °C, maximum value in methane conversion at 0, 1 and 2 % of CO_2 was 69.0, 70.1 and 60.5, respectively. Existence of 1 % of CO_2 enhanced methane conversion in comparison with methane with no CO_2, suggesting that small amount of CO_2 can enhance reaction as follows: $CH_4 + CO_2 \rightarrow 2H_2 + 2CO$. The decrease from the maximum value of methane conversion to the value at 6 h after start of reaction at 0, 1 and 2 % of CO_2 were 17.8, 8.4 and 0.7 %, respectively. This result suggests that existence of CO_2 can inhibit the decrease in methane conversion. There is provability that presence of CO_2 may prolong catalytic lifetime.

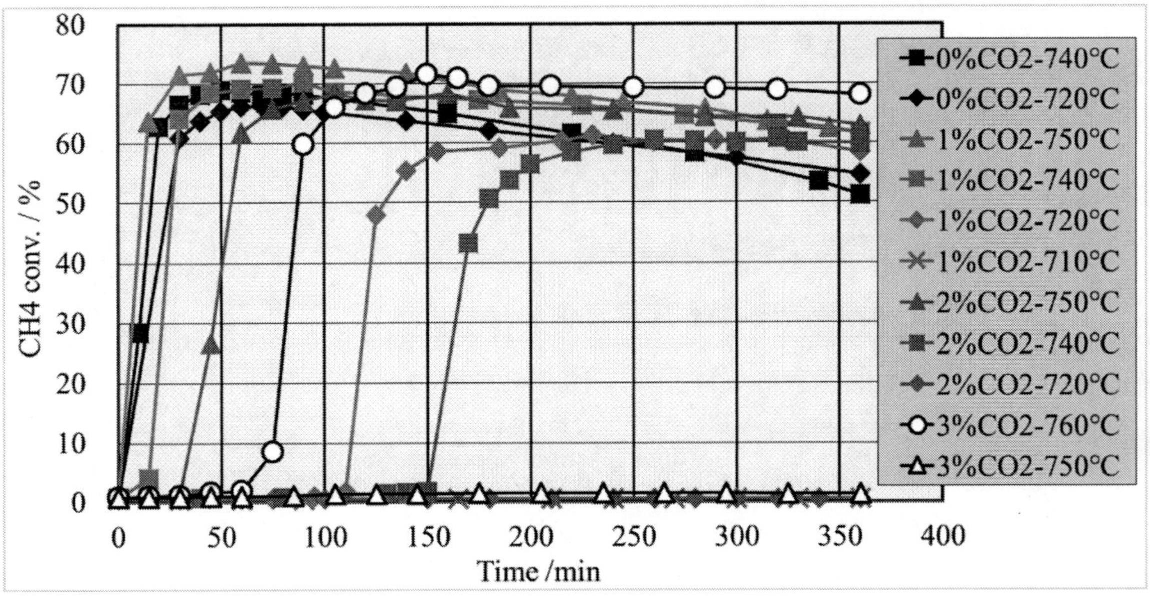

Fig. 5 Change of methane conversion with time-on-stream.
Flow rate of CO_2-containing methane = 20 ml/min.

Conclusions

Methane decomposition reaction using Fe-supported alumina catalyst was investigated to produce hydrogen and carbon. Higher reaction temperature led to shorter induction period before methane conversion starts and slightly higher maximum in methane conversion, but led to larger decrease in methane conversion. Higher flow rate of methane did not affect the maximum in methane conversion rate so significantly, suggesting that this reaction reached equilibrium state. Higher flow rate led to larger decrease in methane conversion, shorter catalytic lifetime and less amount of deposited carbon. Moreover, effect of carbon dioxide with methane on catalytic performance was investigated, since biogas which can be obtained by fermentation of biomass contains not only methane but also carbon dioxide. The presence of carbon dioxide in methane gas inhibited reduction of iron oxide, and as a result, longer induction period and higher reaction temperature were necessary for methane conversion.

References

1) M. Inaba, K. Murata, M. Saito, I. Takahara, N. Mimura, *React. Kinet. Catal. Lett.,* **77,** 109 (2002).

2) K. Murata, M. Inaba, M. Saito, I. Takahara, N. Mimura., *J. Jpn. Petrol. Inst.,* **46,** 196 (2003).

3) K. Murata, M. Inaba, M. Saito, I. Takahara, N. Mimura., *React. Kinet. Catal. Lett.,* **80,** 39 (2003).

480389 Calcium Removal for the Opportunity Crude Using the Deritives of Pyrurvic Acid

Abstract:

Due to the growth limitation of light crude oil production, refinery companies are showing interest in heavy crude oil containing the high contents of S, Ca, organic acids, and metals. The market price of heavy crude oil is much lower than that of light crude oil and the discount rate depend on the impurity contents (averagely 10 ~ 25%). Thus, as increasing the handling capacity of heavy crude oil, economic feasibility improved.

Ca is one of the major impurities in heavy oils, occurring several problems during the refining processes such as fouling in pipe systems and heat exchangers and the deactivation of catalyst. Calcium naphthenate (CaNA), combined with the Ca cation and the anion of naphthenic acids, is difficult to remove because of complex and various molecular structures of naphthenic acids.

In this study, calcium removal from opportunity crude oil was conducted using the deritives of pyruvic acid. The effect of operating variables on the Ca removal efficiency was experimentally studied using Bottle test method. The Ca concentration and water content in the oil samples were analysed using inductively coupled plasma (ICP) and Karl-Fisher method, respectively.

481455 Characteristics of the Hydrocracked Products and Asphaltene with Conversion in a Vacuum Residue Hydrocracking with MoS_2 Dispersed Catalyst

Abstract:

As one of the promising technologies for upgrading heavy oils, slurry phase hydrocracking with dispersed catalysts has increased interest in recent times. However, the formation of sediments and coke at a higher conversion is still known as a problem. This study is to find the characteristics of the hydrocracked products and asphaltene with a change of conversion during the hydrocracking of vacuum residue with molybdenum precursor. The experiment was carried out with a variation of reaction temperature from 385 to 440 'C, reaction time from 0.5 to 4.0 hours and catalyst concentration from 500 to 10,000 wt.ppm in a 250 cc batch reactor at an initial hydrogen pressure of 80 bar.

As results, API of liquid product increases but the liquid yield decreases with increasing conversion due to the higher formation of gas and coke. Compared to the reaction at a lower temperature with longer time, the one at a higher temperature with short time showed a slightly better liquid yield, API improvement and MCR removal keeping the low hydrogen consumption but the sulfur and metal removals are inferior relatively at the same conversion. When the concentration of catalyst increases, the improvement of conversion was prominent at the low temperature. API increment and desulfurization was higher than that at low concentration of catalysts. Asphaltene in the liquid phase was analyzed in terms of colloidal stability so that it was shown to be worse with increasing the conversion. Based on the ratio of iC4 to nC4 in the gas products, the characteristics between thermal and catalytic cracking was compared with the change of conversion.

478844 Purification of Styrene from the Styrene/Ethylbenzene Mixture By a New Technique Combining Distillation and Crystallization

Abstract:

Purification of styrene from the styrene/ethylbenzene mixture by a new technique combining distillation and crystallization

Lie-Ding Shiau[1,2]

[1]Department of Chemical and Materials Engineering,

Chang Gung University

Taoyuan, Taiwan, R.O.C.

[2]Department of Urology,

Chang Gung Memorial Hospital

Linko, Taiwan, R.O.C.

E-mail:shiau@mail.cgu.edu.tw

Abstract

Stripping crystallization (SC) is a new separation technique which combines distillation and crystallization. Basically, SC is operated at a triple-point condition, in which the liquid is simultaneously vaporized and crystallized due to the three-phase equilibrium. In essence, the process is continued until the liquid phase is completely eliminated and only the pure solid crystals remain in the feed. Thus, crystal washing is not required since only pure solid crystals remain in the feed and no impurities are adhered on the crystal surfaces at the end of SC. In the previous research, SC has been successfully applied to separate p-xylene from the mixed xylenes by Shiau et al. (2005, 2008, 2013).

SC is introduced in this work to purify styrene (ST) from the binary system consisting of ST and ethylbenzene (EB). A thermodynamic model is developed to determine the three-phase equilibrium conditions and direct the SC experiments. The experiments show that SC can be effectively applied to purify ST from the mixture. However, as EB can be incorporated into the ST crystal lattice due to the similar structure between these two compounds, it becomes more difficult to further purify ST for a higher concentration of EB in the mixture.

References:

1. Shiau, L. D.*, Wen, C. C. and Lin, B. S. "Separation and purification of p-xylene from the mixture of m-xylene and p-xylene by distillative freezing", *Ind. Eng. Chem. Res*, **44(7)**, p2258-2265, 2005.

2. Shiau, L. D.*, Wen, C. C. and Lin, B. S. "Separation of p-xylene from the multicomponent xylene system by stripping crystallization", *AIChE Journal*, **54(1)**, p337-342, 2008.

3. Shiau, L. D.*; Liu, K. F. "Investigations into the effects of the cooling rate on stripping crystallization", *Ind. Eng. Chem. Res*, **52**, p1716-1722, 2013.

Keywords: crystallization; distillation; purification; styrene

CARBON NANOTUBES FOR ENVIRONMENTAL APPLICATIONS: SOLID-PHASE EXTRACTION AND GAS SENSING FOR CHLORINATED PHENOLIC COMPOUNDS

Walter Den[1], Nurak Gridanurak[2], Ekkachai Kanchanatipk[3] and Sarttrawut Tulaphol[2]
(1) Environmental Science and Engineering, Tunghai University, Taichung, Taiwan, (2) Thammasat University, Bangkok, Thailand, (3) Chulalongkorn University, Bangkok, Thailand

Introduction

Polychlorinated-p-dioxins (PCDDs) and polychlorinated furans (PCDFs) have been identified as the most toxic substance persistent in the environment, released mostly from incineration processes. The formation of these toxic substances is mainly caused by the rearrangement of chlorinated aromatic compounds or precursors to form the dioxins and furans in the post combustion region of incineration system [1]. One of the predominant precursors for the formation PCDD/Fs is chlorinated phenolic compounds [2]. US Environmental Protection Agency (EPA) listed PCDD/Fs as a priority pollutant due to the highly toxic in small amount and classified as a human carcinogenicity and mutagenicity [3]. Therefore, the quantity of PCDD/Fs in the incinerator must be thoroughly analyst before it is emitted into the atmosphere.

A common analytical technique employed for PCDD/Fs with trace concentration in gaseous phase is solid-phase micro extraction (SPME). SPME is a traditional pre-concentration technique based on an adsorption-desorption technique coupled with analytical devices for the quantification of chemicals. Additionally, various types of SPME adsorbents have been applied such as titania nano-particles [4], activated carbons [5], graphene [6], and carbon nanotubes (CNTs) [7].

The properties of CNTs as a sorbent material have been studied by a number of researches. Because CNTs provide high values of absorptivity, uniform porosity, high thermal and chemical stability, and can be improved the performance by incorporate with variety of materials [8]. As mentioned earlier, the application of SPME technique performs mainly by sorption process. Hence, the study of adsorption behavior should be remarked as priority task in order to understand the attraction of adsorbate-adsorbent. There have been several investigations conclusively shown that the adsorption occurred by the π-π interaction among the adsorbate-adsorbent and the adsorption ability is significantly effect by physical properties of adsorbent including surface area and porous structure [9]. For example, Hsu and co-workers [10] performed a series of experiments to determine the adsorption ability of chlorine atoms in six chlorinated volatile organic compounds (carbon tetrachloride, chloroform, dichloromethane, tetrachloroethylene, trichloroethylene, and chlorobenzene) adsorbed onto activated carbon. The adsorption capacity was directly affected by the number of chlorine atoms due to chemical polarity. In addition, the adsorption behavior was described by Langmuir isotherm in their studies.

The affinity of CNTs with aromatic ring compounds makes them suitable to act as the core element of both gas sensing and extraction purposes. The nanostructure of CNTs also provide a significant high surface-to-volume ratio advantageous for high-responsive gas sensors. The adsorption of gas on the CNT surface at room temperature varies its electrical properties with fast response, which enables them to be a potential candidate for gas sensing applications. For example, Chopra and co-workers [11] developed a method using wired SWNTs to study the resonant frequency shift phenomenon for several polar and nonpolar gaseous compounds (He, Ar, N_2, O_2, CO, NH_3), and found that there existed a correlation between the dielectric constant of SWNT in contact with these gaseous compounds and the responding resonant frequency shift, and that these correlations may be used for gas detection in the parts-per-million (ppm) level. Furthermore, for the detection of organic vapors, Consales and co-workers [12] applied Langmuir-Blodgett method to prepare SWNT on standard silica optical fibers (SOFs) and quartz crystal microbalance (QCM) elements. They tested the changes in frequency shift and optoelectronic signals during adsorption of volatile organic compounds (VOCs) such as toluene and xylene, and estimated sensitivity could reach parts-per-billion levels.

CNTs have also been applied for solid-phase extraction (SPE) and micro-extraction (SPME) techniques. Polyaniline multiwalled carbon nanotubes (PANI/MWCNTs) composition, coated on platinum wire as SPME coated fiber, was applied to concentrate palmitic and oleic acids. The results show that PANI/MWCNTs coated SPME probes obtained stronger signal in GC-MS chromatograms than PANI coated probes [7]. Furthermore, carbon nanotubes-reinforced hollow fiber solid-phase microextraction (CNTs-HF-SPME) connected with HPLC-photodiode array detection was applied to determine five carbamate pesticides in apples. This method can improve the detection limits from (0.09-6.00 ng/g) [13]. The results demonstrated that CNTs could develop performance in term of pretreatment and procedure for the determination of trace carbamate pesticides in apples.

This paper presents the results of our work to apply CNTs for both gas sensing element and extraction media, using a family of chlorinated phenols as the target compounds. The CNTs were made into either a layer on silica micro-particles for packing and extraction purpose, or into the form of a membrane sheet for gas sensing purpose. The adsorption characteristics of chlorinated phenolic compounds captured by synthesized MWCNTs were also characterized and reported herein.

Methodologies

Carbon Nanotubes Synthesis

Floating-catalyst chemical vapor deposition (FCCVD) is used for the growth of CNTs and their embedment on silicon wafer. The system includes a set of concentric quartz tubes, a tubular furnace, and a set of chiller. The inlet to the reactor tube connects to the mass flow rate controllers, which control the flow of hydrogen, argon and methane individually.

Before the furnace is turned on, Ar is introduced into the system at a flow rate of 200 sccm for 10 min to eliminate the air in the tube. Then the second stage is heated to the desired target temperature at a heating rate of 10°C/min from room temperature. At the same time H_2 gas is introduced into the system at a flow rate of 100 sccm with a constant Ar flow rate. Ferrocene is vaporized at 200°C in the first stage zone and the vapor is carried by gas flow into the second stage of the furnace. When the temperature of the second stage reaches the desired temperature, methane is introduced to the reactor with the flow rate of 100 sccm. The reaction time for CNTs growth is carried out for 30 minutes, then the CH_4 gas is turned off and system is cooled down to room temperature under H_2 and Ar gases.

Surface Modification of CNT

500 mg of CNTs were added into a flask containing 120 ml of 69% HNO_3 and 95% H_2SO_4 mixture. The HNO_3/H_2SO_4 mixture was in the ratio of 1:3 by volume. The CNTs were refluxed in the acid mixture at 70°C for 12 h to attach carboxylic functional group (COOH) on the surface of the CNTs. Then the acid functionalized CNTs were filtered through a 0.45 μm pore-sized PVDF membrane and washed with DI water repeatedly until the pH reached neutral. The functionalized CNTs were dried in a vacuum oven at 60°C for 12 h.

CNT Membrane Preparation

Fifty mg of functionalized CNTs were dispersed in 150 ml deionized water using 30 W ultrasonic for 15 min. Then 1 ml of a surfactant, and 100 ml DI water was added into the above suspension of CNTs. The mixture was further sonicated at 30 W for 15 min. The suspension of well-dispersed CNTs was filtered by a vacuum pump through a 0.45 μm Durapore PVDF membrane filter. The filter was placed in air at room temperature for 1 h and then dried in a vacuum oven at 60°C for 3 h. The CNT membrane was peeled off from the filter and kept in a desiccator until application.

Chlorinated Phenols Adsorption

MWCNTs embedded on SiO_2 particles were used as adsorbent and phenolic compounds (phenol, 2-chlorophenol, and 2,4-dichlorophenol) were used as adsorbate in gas adsorption experiments. 5 mg of adsorbent was placed in the certain volume of sample container (55ml) with a septum or sampling point on the cap. The MWCNTs was purged by N_2 to eliminate air in the system. Then, chlorinated phenolic compound was injected into the container under constant temperature above its melting point in order to prepare the concentration (1-500 ppm). It was further evaporated by heating temperature at 80°C in the oven for an hour. Afterward, the adsorption was taken place for overnight to ensure the system reached equilibrium at room temperature. Furthermore, SiO_2 particles were tested in the blank experiment. The equilibrium concentrations of each gas were collected using syringe inserting through the septum on the top of bottle, then injected to analyze by gas chromatography–mass spectrometry (GC-MS, Thermo Scientific, USA) from time to time.

A GC system equipped with a TG-5MS capillary column (length 30 m, i.d. 0.25 mm and film thickness 0.25 μm). The GC-MS system was set up with the injection temperature at 280°C and applied helium as a carrier gas with 1 ml/min of flow rate. In addition, the split flow was controlled at 10 ml/min. The transferring line between GC and MS system was maintained at 250°C. The GC-MS system operated at a programmed oven temperature from 70°C (held for a min) to 110°C with 20°C/min of heating rate, then heated to 200°C with 25°C/min of heating rate, thereafter remained at 200°C for 2 min. The ion source temperature was maintained at 200°C, and full scan electron ionization mode operated in the MS detector where data in range of *m/z* 50-400 were obtained.

Adsorption Behavior Study

Thermogravimetric/differential thermal analysis (TG/DTA, SII Exstar6000, Japan) was performed to determine the activation energy for desorption (E_d) of phenolic compounds adsorbed onto MWCNTs embedded on SiO_2 microspheres. The preparation of saturated adsorbents was conducted by placing 10 g of phenolic compound and 0.2 g of adsorbent in the desiccator containing 10 g of molecular sieve 3A for capturing moisture. After that, the system was carried out for a week in normal condition. Then, the saturated adsorbents were characterized by TG/DTG with different heating rates (1, 3, 5 and 7°C min^{-1}). The thermogram was used to determine E_d for each chemical.

Gas Sensing Measurement

A four-point probe measurement device was adopted for monitoring the change of resistivity of the CNT based gas sensor while exposing to air contaminated with a pre-determined concentration of target gas in the test chamber. A gas sensor was attached to a holder and the four point probe head was placed properly on the surface of the sensor. Through Hall's effect, electrical current passes through the two contact points into the CNT sample. The two other contacts will then measure the resistivity of the sample, and the sensitivity to the change of resistivity is calculated according to the circuit design.

The target gas vapor was used to investigate the sensing capability of CNT sensor by measuring the changes in resistance. Air was used as both a reference and a diluting gas. The chamber was cleaned using vacuum pump; air was then fed into the chamber using a pump with the flow rate of 300 ml/min. After the system reached a stable state for a while, a certain amount of the analyte in a Tedlar bag was then fed into the test chamber through mass flow controller.

Solid Phase Extraction

SiO_2 particles embedded with MWCNTs were used as adsorbent to extract chlorophenols. SiO_2 particles embedded with MWCNTs was filled into a stainless steel thermal desorption tube (length 89 mm; o.d. 6.4 mm; i.d. 5 mm) (Perkin-Elmer, Waltham, USA) to examine the effect of packing quantity. The adsorbent was packed in the middle of stainless steel tube, surrounded with glass wools on both side and stainless steel sieve on the outlet side. Prior to use, the tube was purged with high purity helium gas (99.9999%) while heating at 350°C to eliminate impurity for 10 hr. One μL of standard CPC was slowly injected into meddle of purified tubes. The tube was taken place to reach equilibrium for 1 hr. The tube was analyzed by automated thermal desorption (ATD) (ATD-100, Perkin-Elmer, USA) equipped with GC-MS. ATD consists of two stages thermal desorption process. First stage, CPC was thermally desorbed by flowing of helium as carrier gas (30 ml/min) into second stage. To re-concentrate CPC, cold trap (-30°C) was used in second stage. Cold trap was rapid heat-desorbed CPC again up to 300°C with heating rate of 40°C/s and transferred by helium gas into the capillarity column in GC-MS with flow rate as GC-MS condition (1 ml/min). To protect condensation, the transfer line was maintained temperature at 250°C. To improve limit of detection, the inlet split at first stage was turned off during desorption process. To protect over-extended chromatography, outlet split flow rate was set at 8 ml/min.

Results and Discussion

CNT Characterizations

CNTs grown from the Fe catalyst particles from ferrocene were mainly multi-walled CNTs (MWCNTs). The diameters range around 30-50 nm. TEM examination shows that the catalyst particle was rising at the tip of nanotube during its growing process, and that there might be a second and more Fe catalyst particles falling on it. There was a hollow section above the catalyst particle in the tip. This reveals that the nanotube could grow even without a catalyst particle on the tip. Because the catalyst particle is circled with nanotube wall, the carbon source for the further growth of nanotube must not come from the precipitation of Fe catalyst particles in the tip, but from the direct deposition of carbon cluster on the open tip.

High-resolution TEM examination verified the tubular structure of the as-grown MWCNTs possessing 16 walls and the inter-wall spacing around 0.18 nm. The inter-wall spacing was not large enough for phenol molecules (~0.46 nm) and other chlorophenols with greater molecular sizes to access. The results suggested that only the inner cavity and the external nanotube surface are available for chlorophenols adsorption on the synthesized MWCNTs.

Raman spectroscopy shows the spectra from 1000-2000 cm^{-1} of the as-synthesized CNTs via FCCVD and the acid-functionalized CNTs. The Raman spectra contain two domains in the range. The first band located at 1345 cm^{-1} is denoted as D band; it is activated by the presence of disorder in carbon system. The second band at 1580 cm^{-1} (G band) is assigned to the in-plane vibration of C-C bond with a shoulder at around 1600 cm^{-1}. D band and G band of the spectra confirmed that the synthesized CNTs were structured with multiple layers. The intensity ratio of D- to G-band (I_D/I_G) reflects both purity and the defect density of CNTs. Furthermore, the spectra of the acid-functionalized CNTs exhibited higher intensity of D-band due to the breaking of sidewall C-C bonds and introduction of the oxygenated functional groups onto the sidewall and the end wall of the nanotubes. The ID/IG increased from 0.94 to 0.97 after acid functionalization.

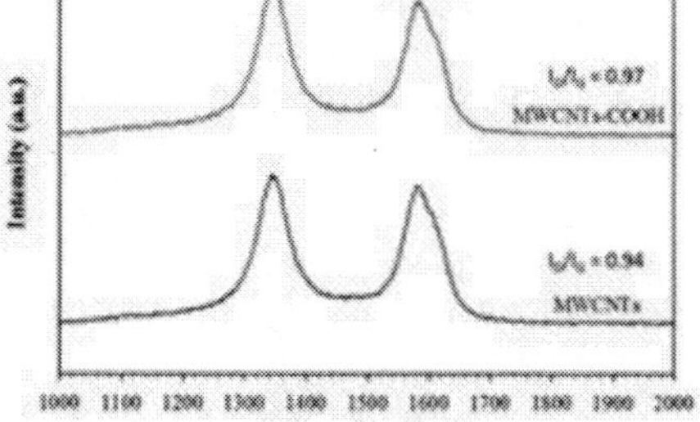

Figure 1. Raman spectra of as-synthesized and acid-functionalized CNTs.

Adsorption Properties of Chlorophenols

To demonstrate attraction behavior of sorbate-sorbent, thermogravimetric /differential thermal analysis (TG/DTG) was performed to identify the activation energy for desorption of phenolic compounds adsorbed onto MWCNTs. Then, the saturated adsorbents were characterized by TG/DTG with different heating rates (1, 3, 5 and 7°C/min). Figure 2 reveal a typical profile of the temperature programed desorption analysis for a phenolic compound, plotted between DTG (μg/min) and temperature (°C). The desorption curve is higher when the heating rate is higher due to increasing of desorption rate.

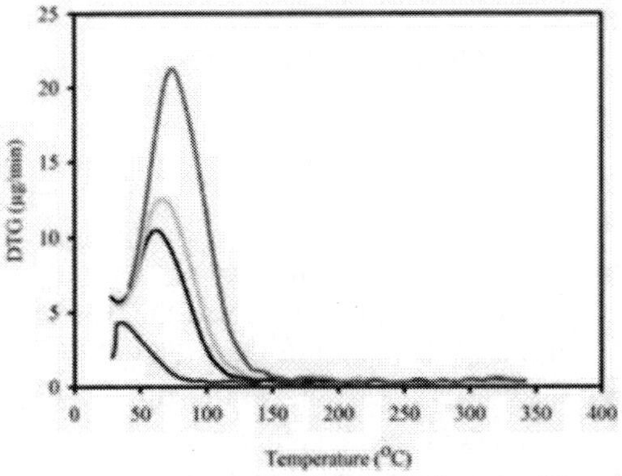

Figure 2. Temperature programmed desorption profile of chlorophenol.

The strength of interactions between chlorinated phenol compounds and CNTs embedded on SiO_2 particles were investigated by desorption energy. Figure 3 shows the plot of a linear function between $2lnT_m$-lnB and$1000/T_m$, it can be revealed that 2,4-dichlorophenol shows the highest value of desorption energy according to the steepest slope. Theoretically, the adsorption behavior can be classified into physisorption and chemisorptions based on the activation energy for desorption (E_d).

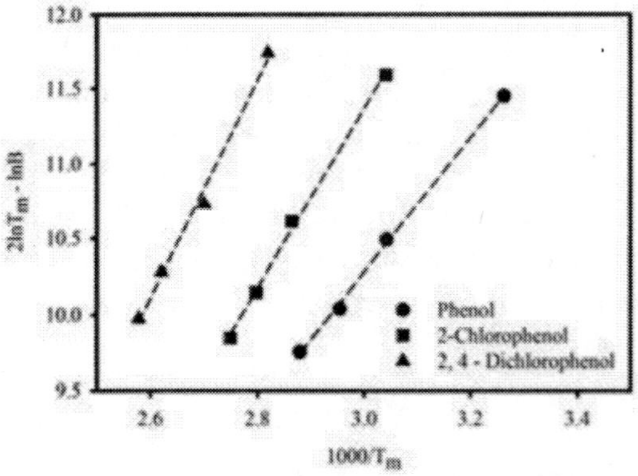

Figure 3. Plot of $2lnT_m$-lnB versus $1000/T_m$ for three (chloro)phenol compounds.

Sensitivity to Chlorophenols

The sensitivity of 100 ppm gaseous chlorophenol on buckypapers is shown in Figure 4. Both buckypapers performed rapid detection when chlorophenol gas was introduced into the chamber. The acid-functionalized CNT buckypaper showed substantially more sensitive than the non-modified one. The carboxylate group on the surface of acid-functionalized CNT could enhance the electron transfer rate, resulting in significantly higher resistance change when exposed to chlorophenol gas. The signal change was about three times higher while the recovery time was slightly longer than that of non-modified CNT buckypaper due to the stronger bonding energy. These results are in good agreement with the results of adsorption behavior test.

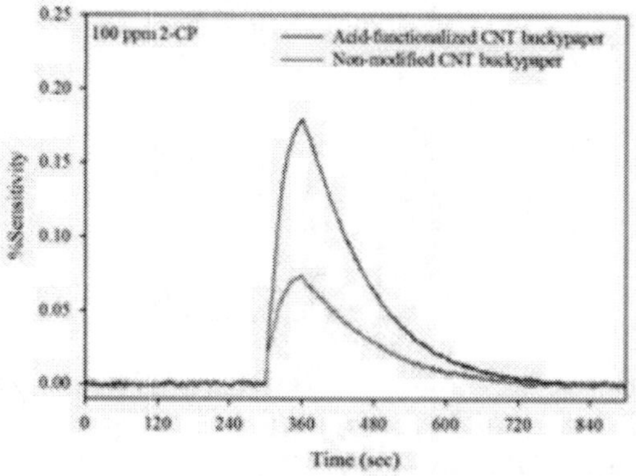

Figure 4. The sensitivity of 100 ppm 2-chlorophenol

The MWCNT gas sensor responses were linear with respect to analyte concentration ($r^2>0.99$), suggesting a first-order interaction kinetics between the analyte and CNTs. From the slope of the lines, it is clear that the sensitivity of CNT buckypaper on 2,4-dichlorophenol was higher than 2-chlorophenol and phenol gases. It is noteworthy that the sensitivity of gas sensor to the target gases increased with an increasing of the number of chlorine atoms, which corresponded with the results in adsorption behavior study.

The sensor was exposed to consecutive pulses of 100 ppm 2-chlorophenol for repeatability test. The exposure duration was 1 min/pulse and then purging with air for 10 min, repeatedly. Figure 5 shows eight reversible cycles of the response curve of the acid-functionalized gas sensor exposed to 2-chlorophenol. The sensor showed insignificant change in response recovery time and the sensitivity. The sensor responded rapidly in few seconds upon the exposure of the target gas.

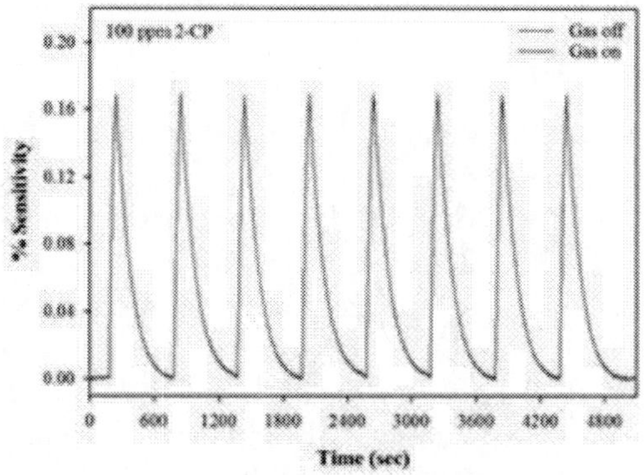

Figure 5. Repeatability test of CNT buckypaper gas sensor for 2-chlorophenol.

SPE Studies

To investigate the performance of MWCNTs embedded in SiO_2 particles using as adsorbent for SPE analysis, commercial adsorbent (Tenex TA and carbopack B) were evaluated to compare the performance. 400 μg/L of (chloro)phenols concentration was used in all experiment. The performance of adsorbent relates with amount of desorbed (chloro)phenols from the tube, displayed as response area. Figure 6 displays comparison of MWCNTs embedded in SiO_2 particles compare against with commercial adsorbents. Tenax TA shows highest response area for P and DCP, which also indicated highest adsorptivity, followed by MWCNTs/SiO_2 and carbopack B, respectively. In the other hand, the response area of MWCNTs/SiO_2 enhanced and resembled with response area of Tenax TA for CP.

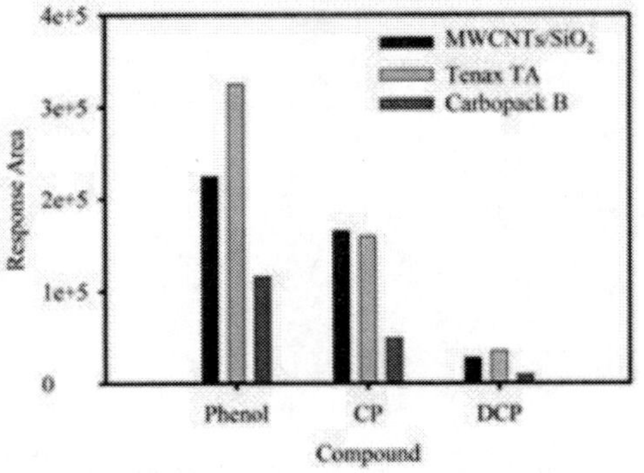

Figure 6. Comparison of MWCNTs embedded in SiO_2 particles using as an adsorbent in SPE analysis compare with commercial adsorbents (Tenax TA and Carbopack B).

Methodologies

(1) HUANG, H. & BUEKENS, A. 1996. De novo synthesis of polychlorinated dibenzo-p-dioxins and dibenzofurans Proposal of a mechanistic scheme, *Science of The Total Environment*, 193, 121-141.

(2) BARANSI, K., DUBOWSKI, Y. & SABBAH, I. 2012. Synergetic effect between photocatalytic degradation and adsorption processes on the removal of phenolic compounds from olive mill wastewater, *Water Res.*, 46, 789-798.

(3) MUKHERJEE, R. & DE, S. 2014. Adsorptive removal of phenolic compounds using cellulose acetate phthalate–alumina nanoparticle mixed matrix membrane, *J. Hazard. Mater.*, 265, 8-19.

(4) LIU, S., XIE, L., ZHENG, J., JIANG, R., ZHU, F., LUAN, T. & OUYANG, G. 2015. Mesoporous TiO_2 nanoparticles for highly sensitive solid-phase microextraction of organochlorine pesticides, *Analy. Chim. Acta,* 878, 109-117

(5) SUN, T., JIA, J., FANG, N. & WANG, Y. 2005. Application of novel activated carbon fiber solid-phase, microextraction to the analysis of chlorinated hydrocarbons in water by gas chromatography–mass spectrometry, *Analy. Chim. Acta*, 530, 33-40

(6) FAN, J., DONG, Z., QI, M., FU, R. & QU, L. 2013. Monolithic graphene fibers for solid-phase microextraction, *J. Chromatogr. A*, 1320, 27-32

(7) KHAJEAMIRI, A. 2012. Preparation of Solid Phase Microextraction (SPME) Probes through Polyaniline Multiwalled Carbon Nanotubes (PANI/MWCNTs) Coating for the Extraction of Palmitic Acid and Oleic Acid in Organic Solvents. *Iran. J. Pharmaceut. Res.* 11.1, 369–374.

(8) SU, F., LU, C. & HU, S. 2010. Adsorption of benzene, toluene, ethylbenzene and p-xylene by NaOCl-oxidized carbon nanotubes, *Coll. Surf. A*, 353, 83-91.

(9) ZHAO, X.S., LU, G.Q. & HU, X. 2001. Organophilicity of MCM-41 adsorbents studies by adsorption and temperature-programmed desorption, *Coll. Surf. A*, 179, 162-269.

(10) XU, M., SUN, Z., CHEN, Q. & TAY, B. K. 2009. Effect of chemical oxidation on the gas sensing properties of multi-walled carbon nanotubes. *Int. J. Nanotechnol.*, 6, 735-744.

(11) CHOPRA, S., MCGUIRE, K., GOTHARD, N., RAO, A. M. & PHAM, A. 2003. Selective gas detection using a carbon nanotube sensor. *Appl. Phys. Lett.*, 83, 2280-2282

(12) CONSALES, M., CRESCITELLI, A., PENZA, M., AVERSA, P., VENERI, P. D., GIORDANO, M. & CUSANO, A. 2009. SWCNT nano-composite optical sensors for VOC and gas trace detection. *Sens. Actuat. B,* 138, 351-361

(13) SONG, X.-Y., SHI, Y.-P. & CHEN, J. 2013. Carbon nanotubes-reinforced hollow fibre solid-phase microextraction coupled with high performance liquid chromatography for the determination of carbamate pesticides in apples. *Food Chem.,* 139, 246-252.

481919 Regression Strategies for Large Data Sets (poster version)

Abstract:

Plant operators and engineers are increasingly using historian data to gain insight into the relationships between and amongst process parameters and product attributes. A universal goal is the determination of the operating conditions that produce the highest product output and/or quality per unit cost. Models provide insight, but seldom account for myriad physical nuances which sometimes exhibit significant influence.

The quantity of data that must be analyzed in pursuit of the sought inferences is invariably large, consisting of hundreds or thousands of quantities sampled at many millions of points in time. It is seldom possible to process data sets of this size in computing resource memory.

A number of frameworks for partitioning large data sets have been developed and popularized, though adoption by industrial companies has been notably slower than by IT-centric enterprises. This paper endeavors to demystify the processing of out-of-memory data for an engineering audience.

Regression, which is fundamental in data analysis, is used to motivate use of the techniques. This paper begins with computing simple statistics. Subsequently, some strategies for matrix manipulations (relevant to both direct and iterative methods) are discussed. Finally, an illustrative example of a large data set regression is presented.

Agenda

- Introduction
- Industry Challenges
- Teradata Approach
- Teradata Solutions
- Teradata differentiators

2 TERADATA

IoT – What's New?

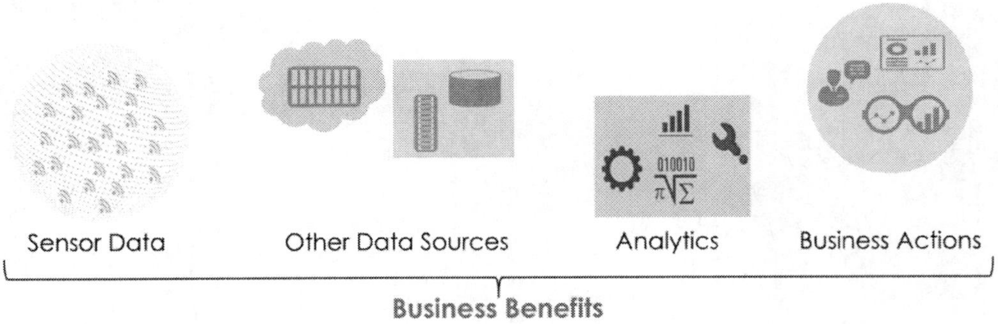

Right data, right insights, right actions driving business outcomes – revenue growth, customer satisfaction, operational efficiency

3 TERADATA.

What are Industry Leaders Saying

- **35%** of US manufacturers are currently collecting and using data generated by smart sensors to enhance manufacturing/operating processes
- Organizations capture but face challenge using data. **48%** spend the most time reviewing data for quality and consistency

Business Focus for IoT initiatives

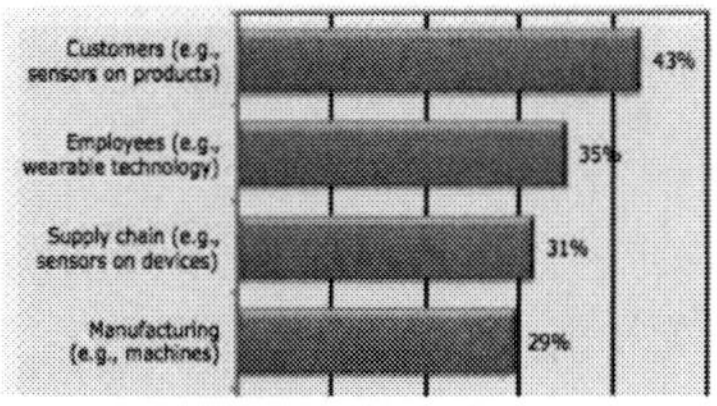

Issues with Implementing Technology

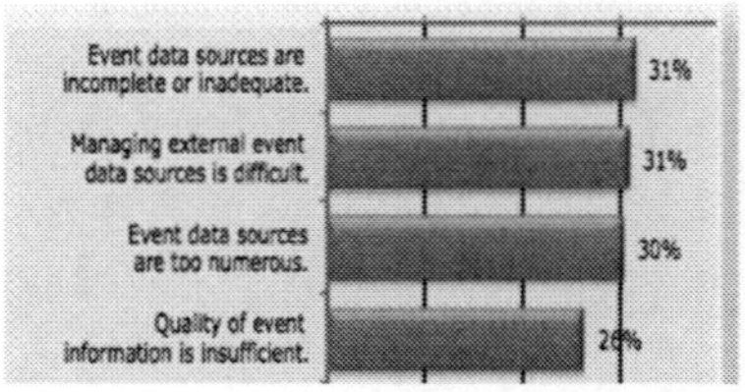

4 TERADATA.

Today's Challenges – Data Issues

It comes from		It can contain…		It has impact
Fleets: from lots and lots of similar things		**Outliers:** Which of my things are behaving differently?	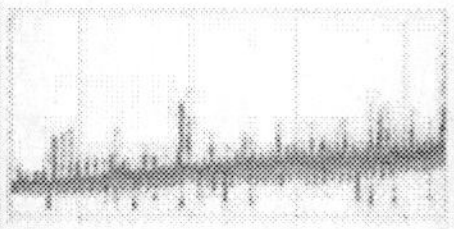	"Fleet-wide" 24/7 for holistic management
Systems: across the same big "thing" or facility		**Emergent behaviour:** Is my system changing to a new state?	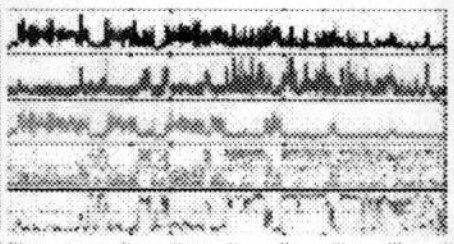	High-level KPIs at business units and facilities level
Collectors: "big models" or monitoring	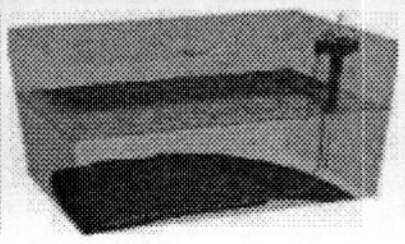	**Events:** are there hidden signals?	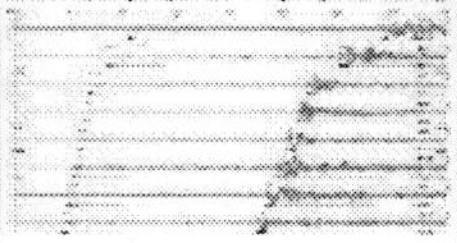	Performed at sub-second level and data kept for decades

5

TERADATA.

Today's Challenges – Analytics Complexity

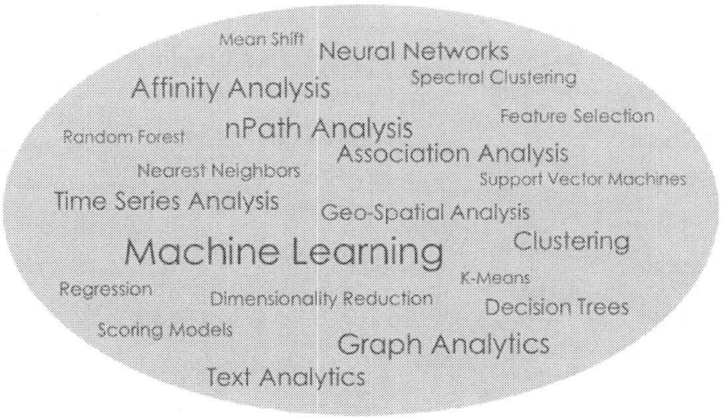

- Analytical Tools
- Model Generation
- Validity
- Accuracy
- Speed
- Scalability
- Repeatability

6

TERADATA.

Today's Challenges - Business link – Operationalizing of Analytics

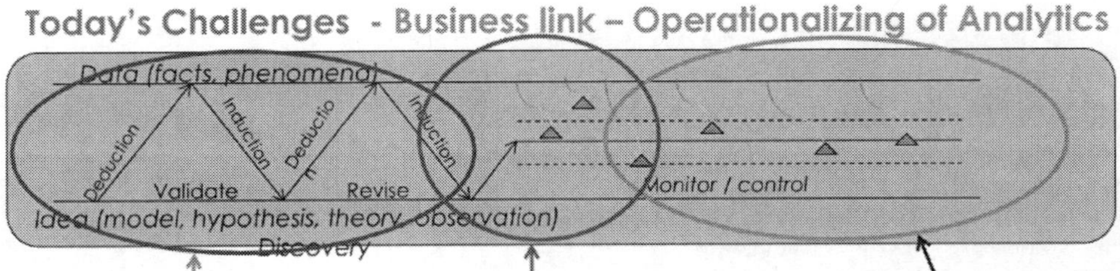

Engineering Focus
- Selecting right tool based on business need
- Setting up a scalable and repeatable environment
- Difficulty in converting business need into an engineering requirement

Implementation
- Missing: Taking discovered relations into production – quickly, efficiently, accurately
- Integration with BI tools, Engineering tools, Reports
- Platform decisions

- Business Contextualization
- Deriving meaningful business actions
- Adaptability of business
- Ability to monitor and measure the impact

7

TERADATA

In Summary, Your Business problem looks like

The Problem

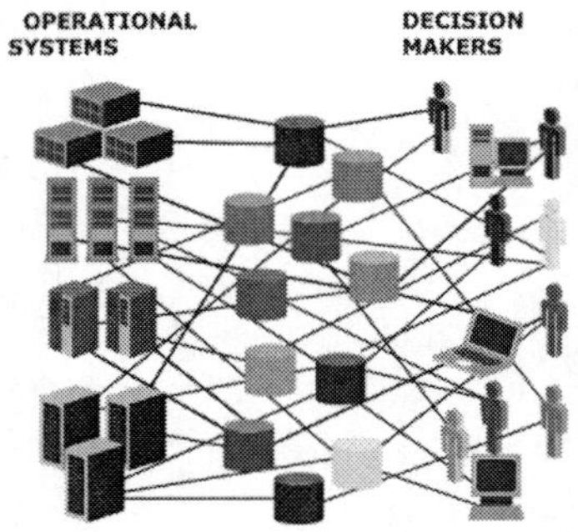

Proliferation of data marts resulted in fragmented data, higher costs, poor decisions

- Represent one silo of the business
- Bringing many silos of data can prove to be difficult
- Cannot keep pace with frequent changes in market conditions
- Decision by leadership is rarely made with complete data picture
- Creates poor communication in value chain
- Data Latency is technology or work process

8

TERADATA

How can we provide decisions at the pace of business

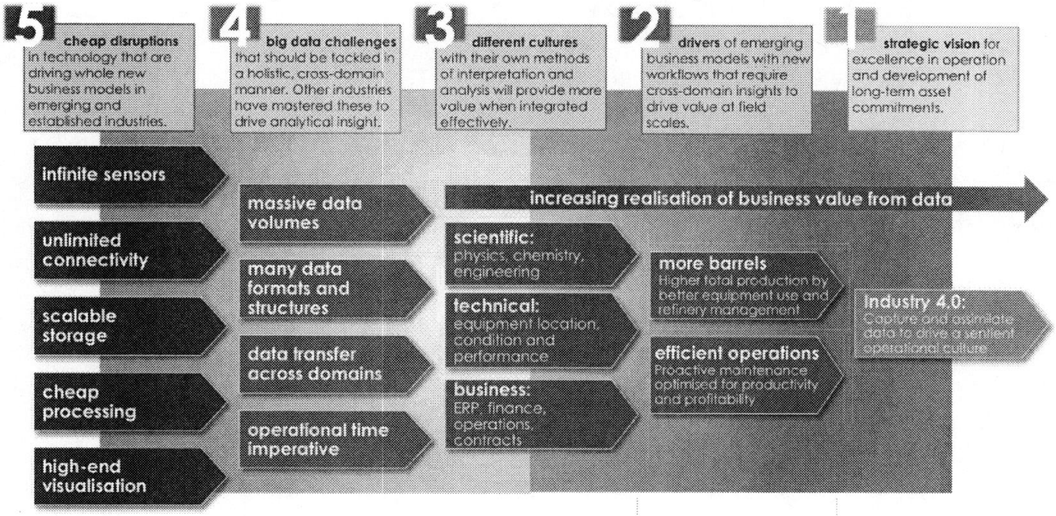

TERADATA

Teradata Approach & Solutions for Industrial IoT

AoT Approach – Overview

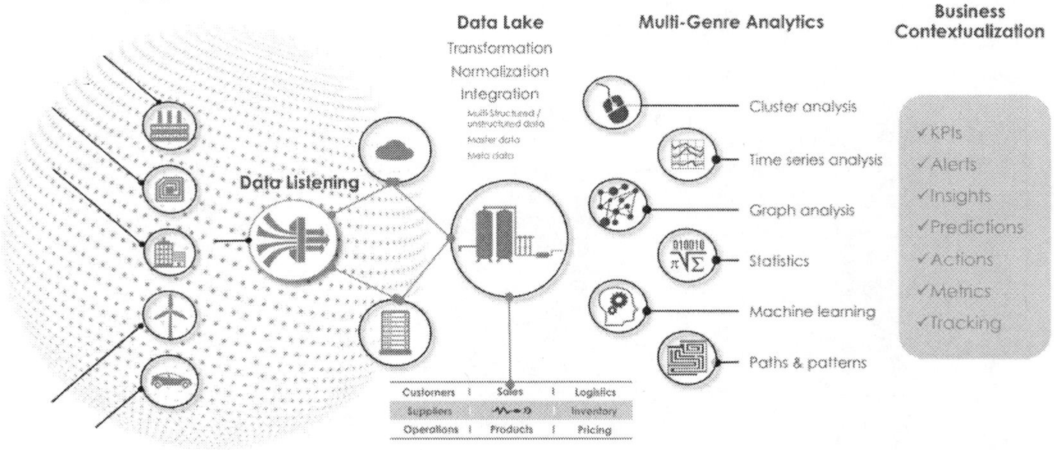

11

TERADATA.

Oil & Gas Applications

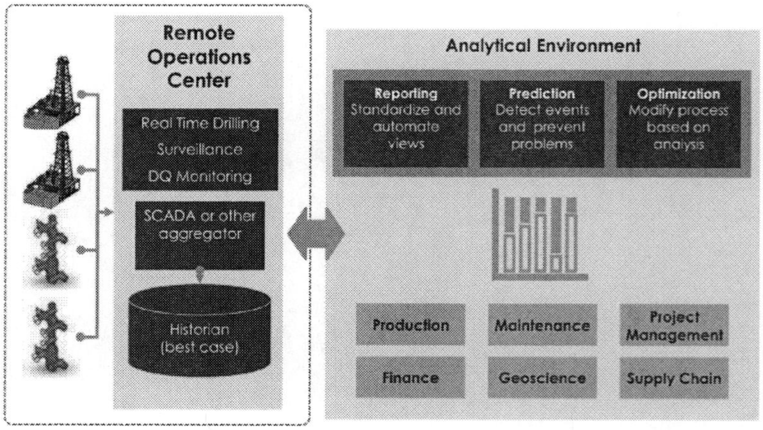

12

TERADATA.

Contents

13 TERADATA.

Solution: Sensor Data Qualification

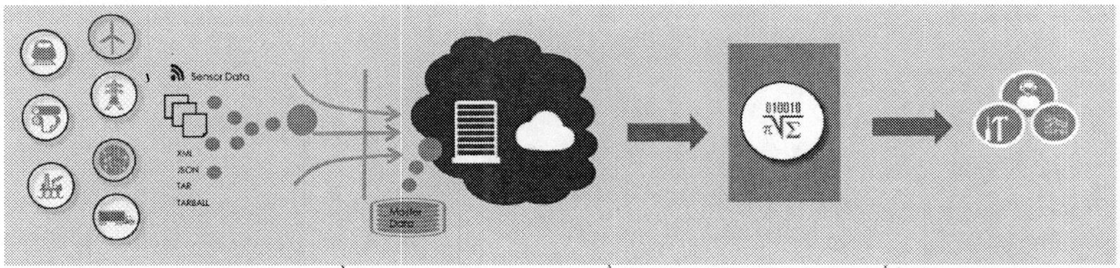

Data Ingestion	Data Transformation & Storage	Data Analytics	Business Contextualization
DQ Stage 1	**DQ Stage 2**	**DQ Stage 3**	**DQ Stage 4**
Qualify Data Ingestion	Qualify Data Transformation	Qualify Data Analytics	Quality Business Relevance

* DQ: Data Qualification

14 TERADATA.

Contents

TERADATA

Solution: Connected Factory

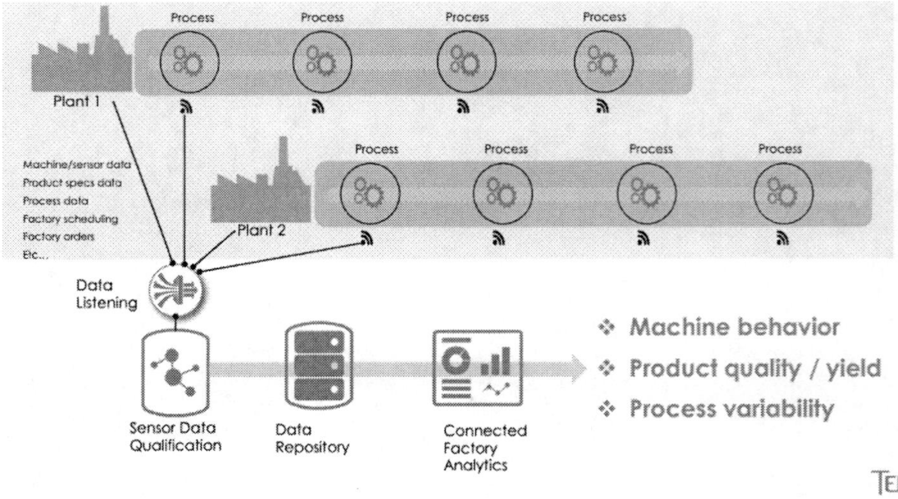

TERADATA

Connected Factory – Machine Behavior

Monitor and analyze machine logs and machine sensors as time series to predict machine failures and prevent unplanned downtime

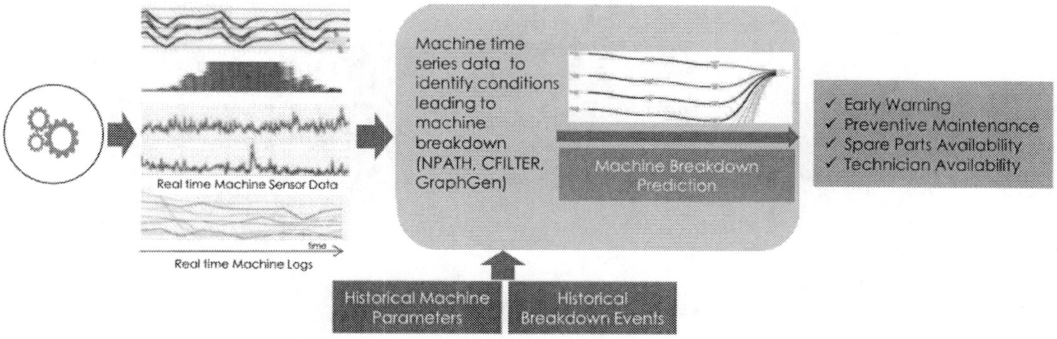

17

TERADATA.

Connected Factory – Product Quality / Yield

Analyze product quality issues to identify conditions that lead to quality defects, yield issues, and build predictive models for detecting these issues

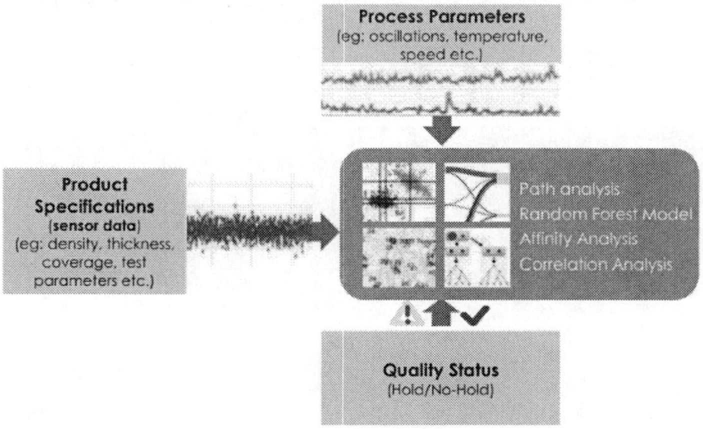

- ✓ Early detection of potential product quality issues
- ✓ Proactive adjustment of conditions that could lead to quality issues, or reduce yield
- ✓ Simulate yield / quality given a set of conditions (design of experiments)

18

TERADATA.

Connected Factory – Process Variability

Analyze process variability impact to address process maturity issues, predict production yields/outputs, and reduce order commitment risk

Sub-Process Variability
(Process flow rules, Real time processing times)

- Variability impact analysis
- What-if analysis & scenario modeling
- Alternate production path analysis
- Order risk analysis
- Raw material availability

- ✓ Increased productivity
- ✓ Faster time to process maturity
- ✓ Fulfillment predictions
- ✓ Risky order alerts
- ✓ Lowered work-in-process inventory
- ✓ Improved material availability

19 TERADATA.

Contents

- Sensor Data Qualification
- Connected Factory
- Smart Assets
- Connected Supply Chain
- Services 360

20 TERADATA.

Solution: Smart Assets

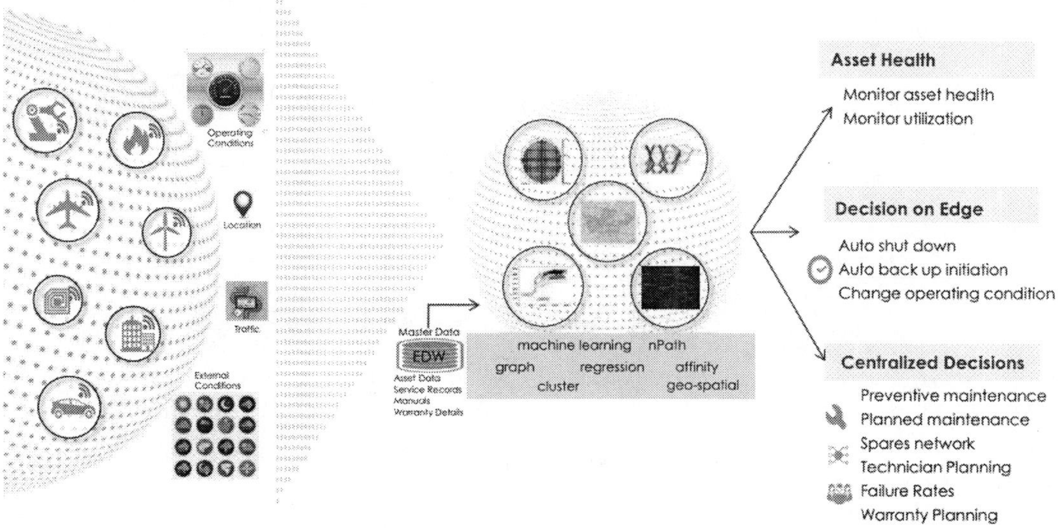

Smart Assets – Use Cases

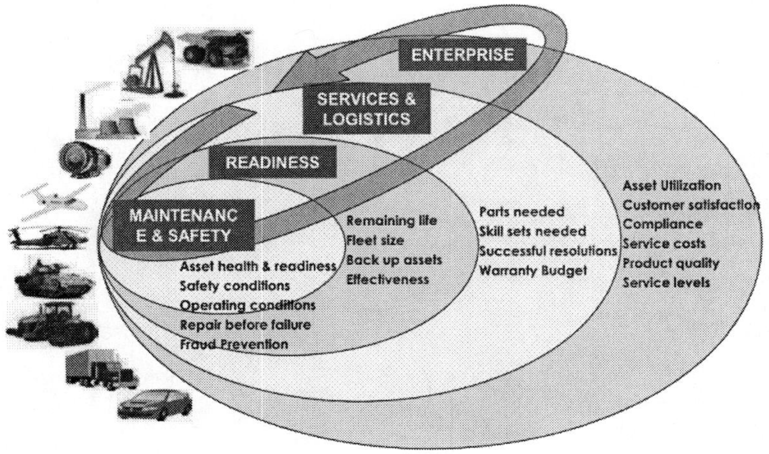

22

TERADATA.

Contents

- Sensor Data Qualification
- Connected Factory
- Smart Assets
- **Connected Supply Chain**
- Services 360

23

TERADATA

Solution: Connected Supply Chain

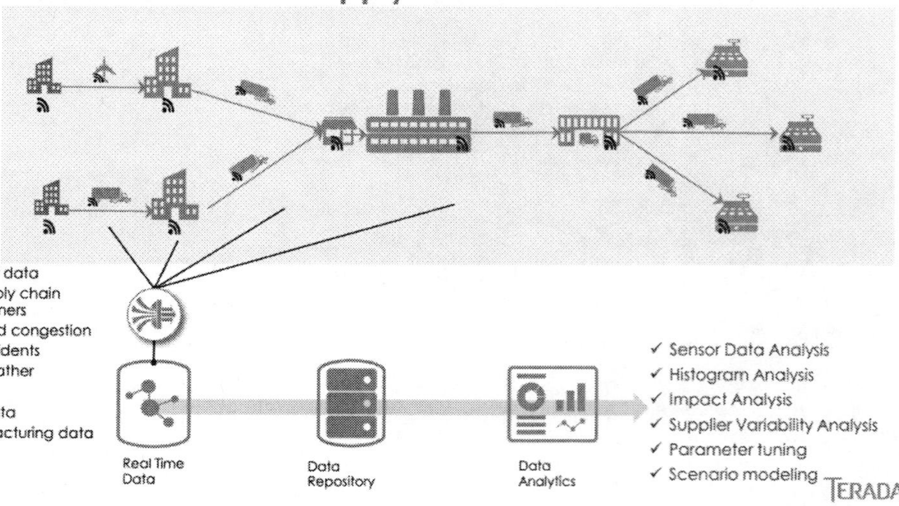

Partner data
- Supply chain partners
- Road congestion
- Accidents
- Weather

GPS data
Manufacturing data

- ✓ Sensor Data Analysis
- ✓ Histogram Analysis
- ✓ Impact Analysis
- ✓ Supplier Variability Analysis
- ✓ Parameter tuning
- ✓ Scenario modeling

24

TERADATA

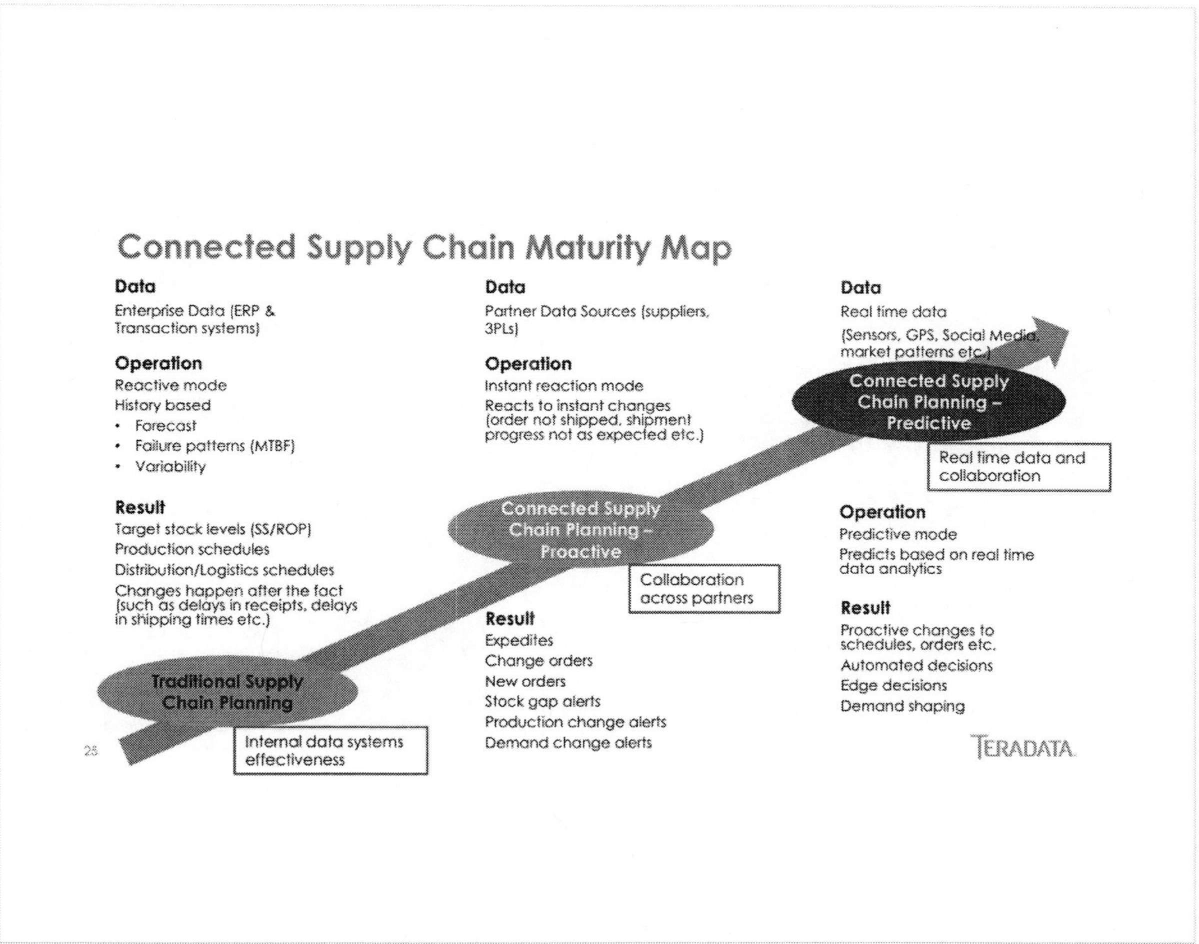
Connected Supply Chain Maturity Map
Data
Enterprise Data (ERP & Transaction systems)
Operation
Reactive mode
History based
• Forecast
• Failure patterns (MTBF)
• Variability
Result
Target stock levels (SS/ROP)
Production schedules
Distribution/Logistics schedules
Changes happen after the fact (such as delays in receipts, delays in shipping times etc.)
Traditional Supply Chain Planning
Internal data systems effectiveness
Data
Partner Data Sources (suppliers, 3PLs)
Operation
Instant reaction mode
Reacts to instant changes (order not shipped, shipment progress not as expected etc.)
Connected Supply Chain Planning – Proactive
Collaboration across partners
Result
Expedites
Change orders
New orders
Stock gap alerts
Production change alerts
Demand change alerts
Data
Real time data
(Sensors, GPS, Social Media, market patterns etc.)
Connected Supply Chain Planning – Predictive
Real time data and collaboration
Operation
Predictive mode
Predicts based on real time data analytics
Result
Proactive changes to schedules, orders etc.
Automated decisions
Edge decisions
Demand shaping
25
TERADATA.

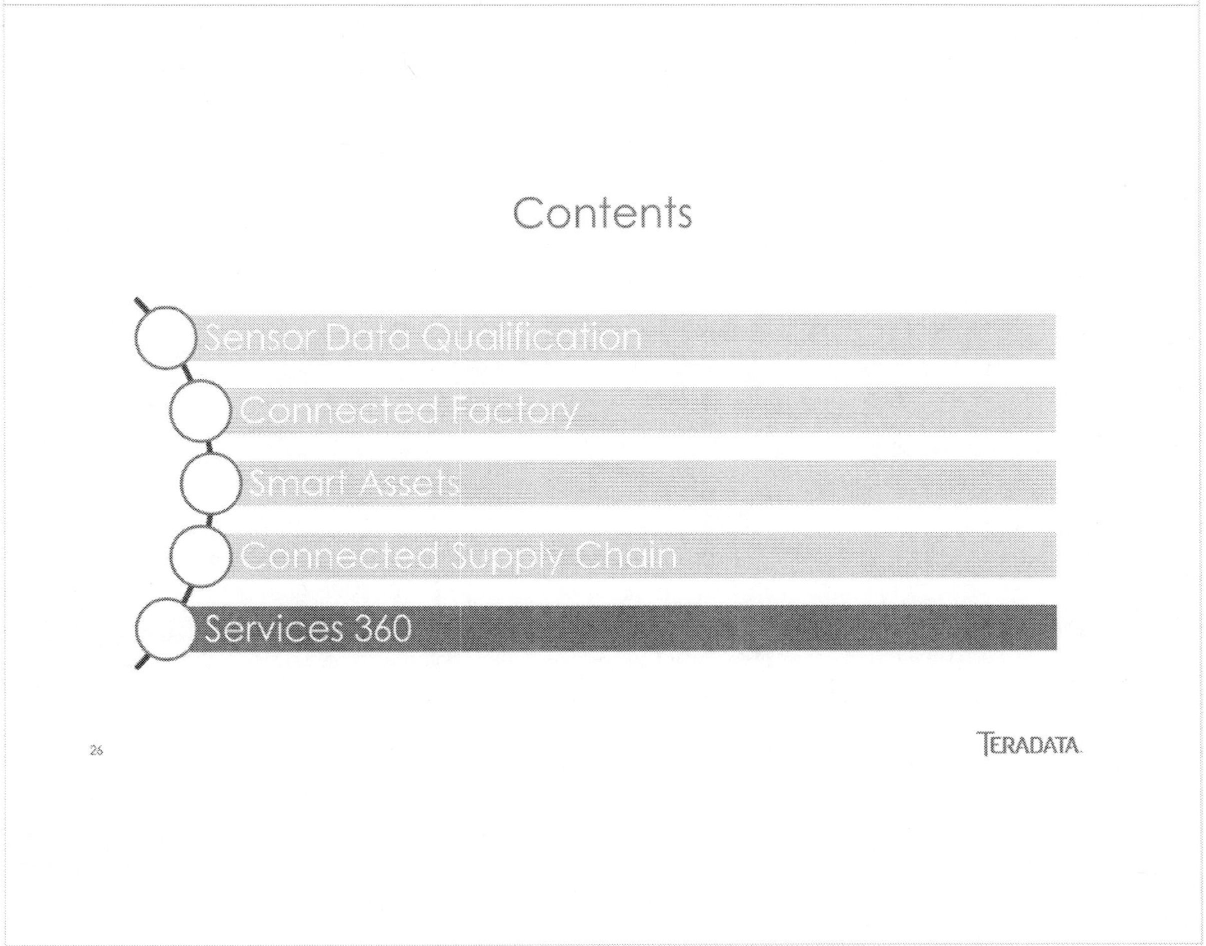
Contents
Sensor Data Qualification
Connected Factory
Smart Assets
Connected Supply Chain
Services 360
26
TERADATA.

Solution: Service 360

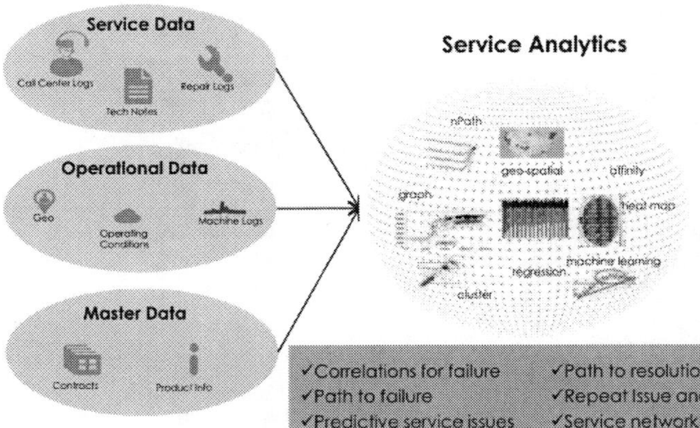

✓Correlations for failure	✓Path to resolutions
✓Path to failure	✓Repeat Issue analysis
✓Predictive service issues	✓Service network analysis
✓Early warning	✓Service profitability analysis

BI/Reporting
✓SLA Adherence
✓Comparative Analysis
✓Customer Patterns

Service Improvements
✓Resolution Guide
✓Product Quality
✓Proactive Fixes

Service Offers
✓Service offer redesign
✓Offer upsell
✓Offer cross-sell

Service Planning
✓Tech Resource Planning
✓Spares Planning
✓Service Network Planning
✓What-if Analysis
✓Capable to Service Analysis

Service 360: Business Benefits

- Proactive issue identification
- Faster & accurate issue resolutions

- Lower cost to serve
- Higher service profitability
- Increase in revenue through cross-sell and up-sell

- Higher net promoter score
- Customer satisfaction & retention

Sample use cases

- Organizations supporting data center temperature control and uptime
- Organizations leasing heavy equipment for production purposes (O&G, Mining)
- Organizations selling enterprise servers with tight support contracts

Enablement

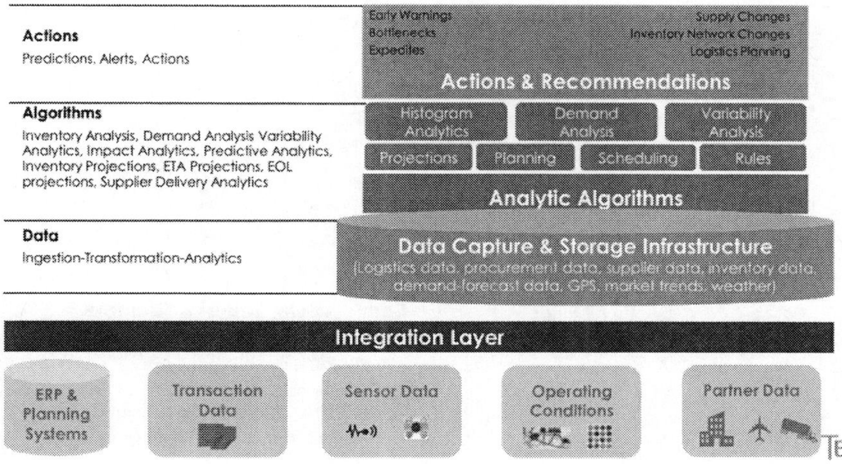

29 TERADATA

Migrate From Projects To Frameworks

Shift focus from a single, tactical exercise to a comprehensive framework

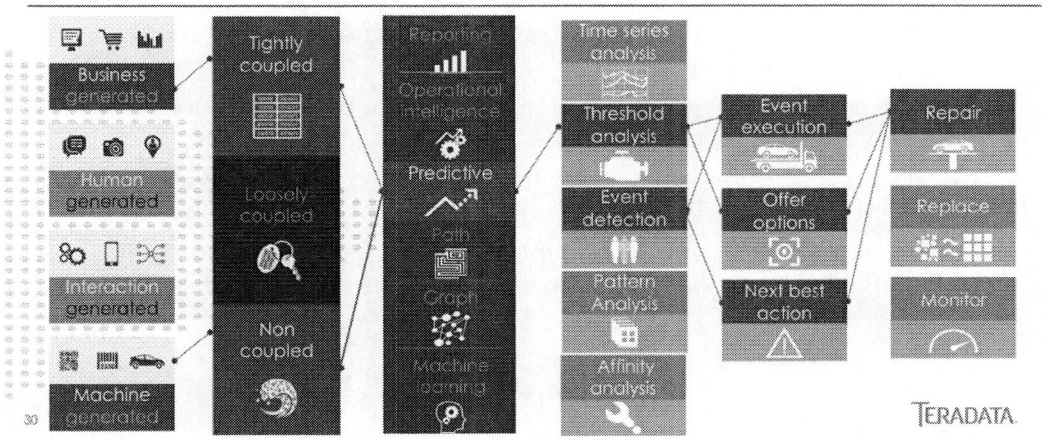

30 TERADATA

TD Differentiators

Outcome based

- ✧ Actionable recommendations
- ✧ Monitors key metrics
- ✧ Incorporates measure, monitor & feedback loops

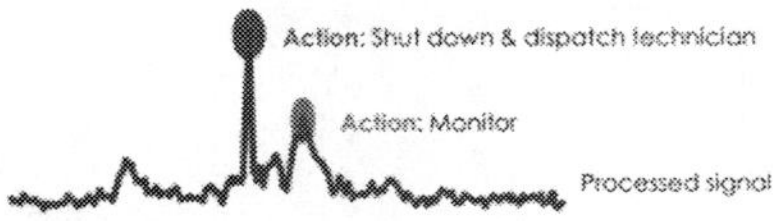

Standardized

- ✧ Specific business problems
- ✧ Proven Approach
- ✧ Data quality management
- ✧ Data models/dictionaries
- ✧ APIs/Analytical Models
- ✧ User Interface

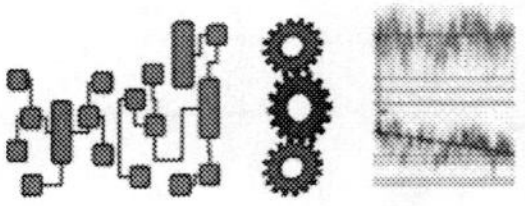

Configurable, Robust & Scalable

- ✧ Business rules can be configured
- ✧ Scalable environment
- ✧ Establishes a flexible environment to solve to other business problems
- ✧ Adapts to multiple technologies
- ✧ Highly Secure

TERADATA.

Concluding Remarks

- You've learnt some new buzz words: Industry 4.0, IoT, Big Data, analytics…and hopefully you understand what they mean and how these ideas impact you
- You've learnt that there is a data driven approach that can build on and enhance your current processes
- You have a sneaking suspicion that it might be possible to unlock more value from your data
- Some of this unlocking could be straightforward integration of data sets but some of it will require some thinking to understand what the data might be telling you (Discovery)
- You have some good ideas on what business value and impact you can drive with this data

32 TERADATA.

481782 The Canopy Platform for Process Monitoring and Optimization (poster version)

Abstract:

Enthought provides product and consulting solutions for the chemical, petrochemical, pharmaceutical, and semiconductor industries. At the core of these solutions is the Canopy Platform, which brings together a host of powerful components for data analysis, visualization, local and remote data access, and data exploration. Canopy Platform solutions allow engineers to easily browse and visualize huge amounts of data stored remotely, do interactive MATLAB-style analysis on it, and develop analysis or detection algorithms and deploy them at scale over large datasets using cluster or cloud compute resources. Detection and prediction algorithms developed in this fashion can be applied to monitor live incoming data and provide feedback to operators, used to provide input to maintenance scheduling, or used to debug production issues. These solutions place a strong emphasis on both convenient desktop user interface (useful for experts to develop their algorithms) and web dashboards for monitoring or data consumption by a broader audience. Web service components makes detection or prediction algorithms written on the platform accessible programmatically, so that the development effort can be reused by other systems and software throughout the business.

In this talk, we show how we leverage the Canopy Platform to build modern, customized software for analyzing plant process systems. We'll demonstrate the ad hoc analysis capabilities and workflow-oriented user interface that simultaneously simplifies open-ended and routine tasks. These capabilities make use of the extensive library of analysis routines that are available to the Canopy Platform and the data management component that enables uniform access to raw files, databases, and remote data stores. We'll walk through a cluster analysis workflow that can make use of a company's existing compute resources to scale to thousands of controllers and years worth of batch or continuous data. We will also describe the larger vision for this software as an automated plant advisor that monitors the status of processes, warns when parameters are out of expected bounds, forecasts failures, and offers solutions to those problems.

481781 Using Analytics for Process Safety (poster version)

Abstract:

Managing plants flooded with thousands non-statistically based alarms daily creates a major challenge for plants wanting to minimize risk. Operators are dismissing non-critical alarms throughout the day, but how many of non-critical alarms does it take to make a critical alarm? 35? 75? 1,000? (is this death by 1,000 cuts?)
Using real-time analytic monitoring can create a scientific way of filtering the data to provide early warning detection, while reducing the noise from fixed set point alarms which may not be relevant for the current process. This presentation will delve into a holistic approach for managing risk within the plant and throughout the whole organization. Monitoring feed stocks, in process parameters, and maintenance data using analytical tests creates an overall view of plant process stability. We will explore case studies and best practices for managing and guiding plant decisions creating a more safe environment for both assets and people.

481780 Making Cognitive Solutions Real Today -Stories and Examples (poster version)

Abstract:

IBM and others have aggressively invested in the advancement of Cognitive technologies such as Natural Language Processing, Machine Learning and Image Recognition, over the past decade . Cognitive driven solution are now having a profound impact on social and economic systems around the world. In this session, IBM will walk the audience through relevant use case and describe what's involved in designing, building and implementing cognitive solutions in a chemical engineering environment. The session will conclude a live demonstration of cognitive solutions.

481779 Practical Challenges and Opportunities in Applying Multivariate Methods in an Online Context (poster version)

Abstract:

The roots of modern multivariate analysis trace their roots back to the 1901 work of Karl Pearson. Despite the well documented benefits of using these techniques for offline analysis, the refining and petrochemical industries have been relatively slow in adopting the technology for online applications. This is surprising considering their enthusiastic and transformational adoption of more computationally intensive tools such as model predictive control.

In this presentation, we describe our experience in using multivariate methods to complement existing approaches and to create novel high value solutions.

480945 Hypothesis Investigation without a Designed Experiment

Abstract:

A designed experiment enables one to infer causal relationships between input variables and the variable of interest. When investigating a hypothesis related to a chemical process issue, a well-designed experiment can provide strong evidence to confirm or refute the claim. What about situations when a designed experiment is not feasible? This situation is common in a chemical manufacturing environment due to the high costs and potential safety concerns associated with recreating the process conditions that caused the original problem. One of Dow's plants faced this challenge as they sought to investigate the claim that the source (supplier) of a raw material was responsible for the lower concentration of active material in the final product. The raw material of interest is highly toxic and has to be manufactured on-site whenever possible to minimize its transport. Occasionally it is obtained from an off-site source when it cannot be produced in sufficient quantities on-site. A combination of visual methods and multivariate modeling techniques were used to examine historical plant data over a period of 3 years. The frequency of the data was hourly (averaged), and approximately three dozen variables were used from multiple sources: online process instruments, online plant analyzers, and the quality assurance lab. The hypothesis was convincingly refuted in the absence of a designed experiment.

480918 Determining Product Specification for Count Data

Abstract:

In order to ensure that a business can produce minimum amount of out of specification product, it is critical to include the statistical knowledge of its process and measurement capability in determining the product specification. This can be done by following "Big Data" concepts, utilizing historical process and/or lab data. As a standard six-sigma methodology, the capability of measurement is commonly evaluated from P/T ratios. However, the evaluation of P/T ignores within-product variation, and is inadequate for non-normal data distributions and/or concentration-dependent standard deviations. We propose a more generally applicable approach, in which we focus on confidence levels (or "critical fallout values") rather than on P/T ratios. Our approach allows for product variation and for all possible data distributions, including discrete distributions.

At The Dow Chemical Company, we have started to apply our generalized approach to the evaluation of method capability in the case of Poisson-distributed data. We will illustrate the merits of our approach from our experiences in this specific application.

480669 EMI Command Center

Abstract:

Dow, being one of the largest and long standing startups (118 years old), accrues immense chemical production process data. It's often said that a journey of a thousand miles starts with one step. This holds true for any effort to harness the power of the data in plant historians, laboratory information management system (LIMS), and online analyzers. Most of our data is underutilized and simply archived in databases and unattainable due to the manner in which data systems were designed and data was stored prior to the age of "Big Data". In order to extract the most significant and immediate value from that data, process design knowledge and operational experience can be used to summarize and aggregate plant data to expose the most relevant information in the right context. At Dow, we want the data to start working for us, rather than us having to work to get the data. The goal of Enterprise Manufacturing Intelligence (EMI) is to make these high value relevant data available in real time to all levels of production management (Operations to Plant and Business Management). Our approach in Manufacturing Analytics is to start small, look for small data opportunities within our plants and operations, and build momentum with success stories. In our quest for total data domination, we are already improving reliability, reducing costs, increasing value, and providing safer operations.

Building a command center for our EMI efforts would mean that we would have all our EMI implementations being monitored through one channel. A command center for EMI would bring the chemical industry to a completely new level of acuity with respect to operations.

481738 DOWN-FLOW Bubble Columns with MICRO-Bubbles for Process Intensification

Abstract:

Down-flow bubble column reactors have received great interest in recent years, due to their small size, stability, high gas-liquid interfacial area concentrations and longer residence times. In conventional bubble columns, average bubble sizes of ~5 mm and void fraction or gas hold up of less than ~20% are usually found leading to a typical interfacial area concentration of 200-300 m2/m3. The major limitation of these conventional bubble column reactors is that they have shorter gas phase residence times due to the high rise velocities of bubbles. Down-flow bubble columns provide longer gas phase residence times and gas-liquid contact times compared to conventional bubble columns. Further intensification can be achieved by efficiently generating micro-bubbles at higher void fractions. High gas-liquid interfacial area leads to high mass transfer rates. In the present work, experiments have been performed in a down-flow bubble column reactor with micro-bubbles generated by a novel mechanism to determine the gas-liquid interfacial area concentrations by measuring the void fraction and bubble size distributions using gamma densitometry and a high speed video camera. The results demonstrate that the new bubble generation technique offers high interfacial area concentrations (1,000 to 4,500 m2/m3) in comparison with previous bubble column reactor designs particularly because of micro-bubbles and high void fractions.

479659 Modeling for the Scale-up of an Ultracapacitor for Energy Storage Applications

Abstract:

Ultracapacitors, also known as supercapacitors, have the potential to meet the high pulse power capability related to the power quality and grid instability for energy storage applications. Ultracapacitors can accept the electric power surges generated from renewable energy sources and deliver the surges of electric power in burst mode to the grid. It is important to calculate accurately the electrical and thermal behaviors of an ultracapacitor for the efficient and reliable systems integration from an application perspective.

In this work, modeling is performed to study the scale-up of an ultracapacitor. A compararison of the experimental discharge-charge curves with the modeling results confirmed that the parameters used to model a 2.7V/650F ultracapacitor could be applied to a large-scale ultracapacitor as 2.7V/1500F and 2.7V/3000F. A three-branch circuit model is employed to calculate the electrical behavior of the ultracapacitor. To predict the thermal behavior of the ultracapacitor, Joule-heat and reversible heat at the electrodes are considered. In addition, the validation of the modeling approach is provided through the comparison of the modeling results with the experimental measurements.

PROCESS INTENSIFICATION IN THE B&P LITTLEFORD PLOW MIXER

Michael A. Smith PE, B&P Littleford LLC, Saginaw, MI

Introduction

Transfer of intermediates from one unit operation to the next takes time, space, and energy—and can be hazardous. Batch processes may be developed and designed to minimize cycle time, capital expense, and operating expense by incorporating process equipment designed to handle multiple unit operations in series. Horizontal plow mixers, such as the FKM and DVT series from B&P Littleford LLC, can handle operations from dry blending and granulation to interface chemical reactions and vacuum drying of dough-like materials.

Several processes are reviewed, each with multiple operations performed sequentially in a Littleford plow mixer. The key capabilities of the Littleford plow mixer are outlined for each unit operation. The impact on capital expense, operating expense, and cycle time for each process is also discussed.

Key Capabilities

The B&P Littleford plow mixer has several design features which enable it to be used in a variety of unit operations. Combined in various ways, these features allow for the development of multi-step processes while minimizing product transfers.

Heat Transfer

Most reaction vessels and distillation units are jacketed, with a heat-transfer fluid such as hot oil, chilled glycol or steam providing heat transfer. The close interaction of the plows with the wall in a Littleford mixer provides a fresh surface for heat transfer throughout the vessel, increasing the overall heat-transfer coefficient for unit operations such as drying.

Full Vacuum to High Pressure Capability

All models of B&P Littleford Mixers are capable of full vacuum. The DVT series adds double mechanical seals and pressure ratings of up to 300 psig for increased flexibility. Steam treatment of materials and pressurized reactions in solvents above their normal boiling point are made possible. Hazardous gases such as anhydrous HCl can be contained and safely reacted.

Granulation

The plow mixer is typically outfitted with one or more high-speed, high-shear "choppers" which act to break up agglomerates into smaller particles. The clearances between the chopper blades, the plows, and the vessel wall provide some limits to this action, along with moderating the speed and duration of the chopping action. Fine powders can often be granulated with the addition of a small amount of liquid, thorough mixing, and limited use of the choppers, resulting in a less-dusty product.

Mechanically-Fluidized Bed

At full speed, the plows are designed to maintain solid particles in a fluidized state, without gas flow or vibratory action. This allows for rapid and complete blending of solids, the efficient distribution of sprayed-in liquid additives, and uniform surface treatment of particles.

Peripheral Equipment

Vacuum systems, dust collection, metered addition systems and feed tanks can all be configured with the plow mixer in different ways, as needed for the particular trial.

Case Studies

The following case studies are actual tests conducted for customers. Confidential information has been withheld or permission given to share results. Each incorporates several unit operations without transferring product from one piece of equipment to another.

Vacuum Drying and Granulation

Many materials are formulated as a stable suspension or slurry, but are preferred to be shipped and handled as a solid. The preferred particle size may vary widely, but a "non-dusty" product presents fewer industrial hygiene concerns, and so a coarser granulation is desirable.

A typical application is seen in Figure 1. This product was temperature-sensitive, and therefore was dried under vacuum. As with many slurries which are dried to a powder, this material went through a cohesive "dough-like" phase, which required significant horsepower. The end of the cohesive phase is marked by a significant reduction in power draw. At this point the chopper was used to break up the product, which could then be more quickly and uniformly dried to the appropriate moisture. Granulation occurred at the same time, as large particles were broken down by the interaction of the chopper with the plow, and fine particles re-agglomerated to a characteristic size, as long as sufficient moisture remained.

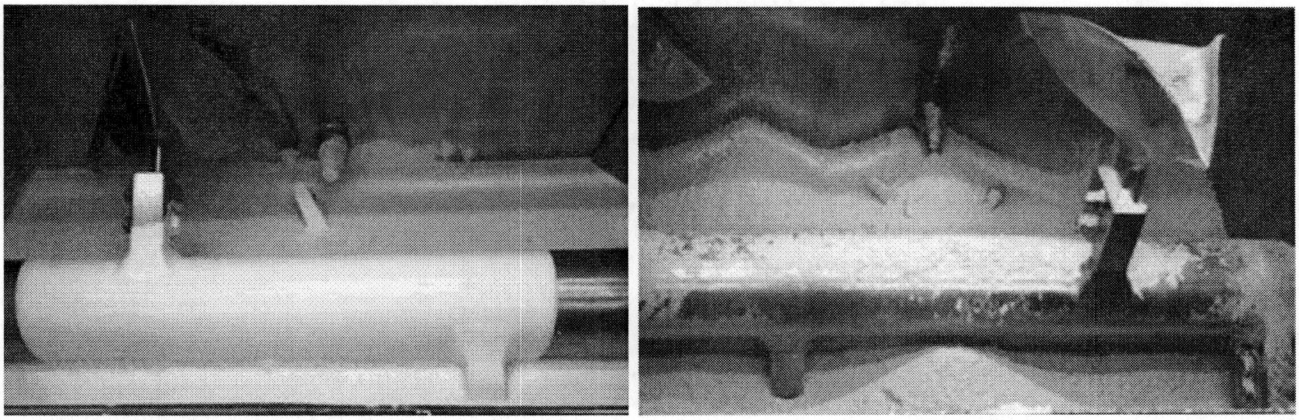

Figure 1: Product before and after vacuum drying and granulation

Limited-Water Reaction

Reaction of a phosphonate acid with ammonium carbonate, to produce the corresponding ammonium phosphonate, can be easily accomplished in a slurry.

$$2\ H_2PO_3\text{-}R\text{-}CO_2H + 3\ (NH_4)_2CO_3 \rightarrow 2\ (NH_4)_2PO_3\text{-}R\text{-}CO_2NH_4 + 3\ CO_2 + 3\ H_2O$$

Since the reaction generates water, only a small quantity of water is actually required to initiate the reaction. A significant reduction in reaction volume can be realized by reacting the two solids as a "wetcake" as long as the wetted solids can be vigorously mixed. Liberation of carbon dioxide gas makes this a net endothermic reaction.

The plow mixer provided the appropriate agitation and heat transfer to convert the acid to the salt quickly while maintaining the mixture at 17 to 20°C. Full vacuum and a 38°C jacket temperature was used to reduce the moisture to less than 6%. Finally a second ammonium additive was blended into the product before discharge. Figure 2 depicts a process flow chart, which shows the five separate unit operations which were performed in sequence, in a single vessel, without transferring the material.

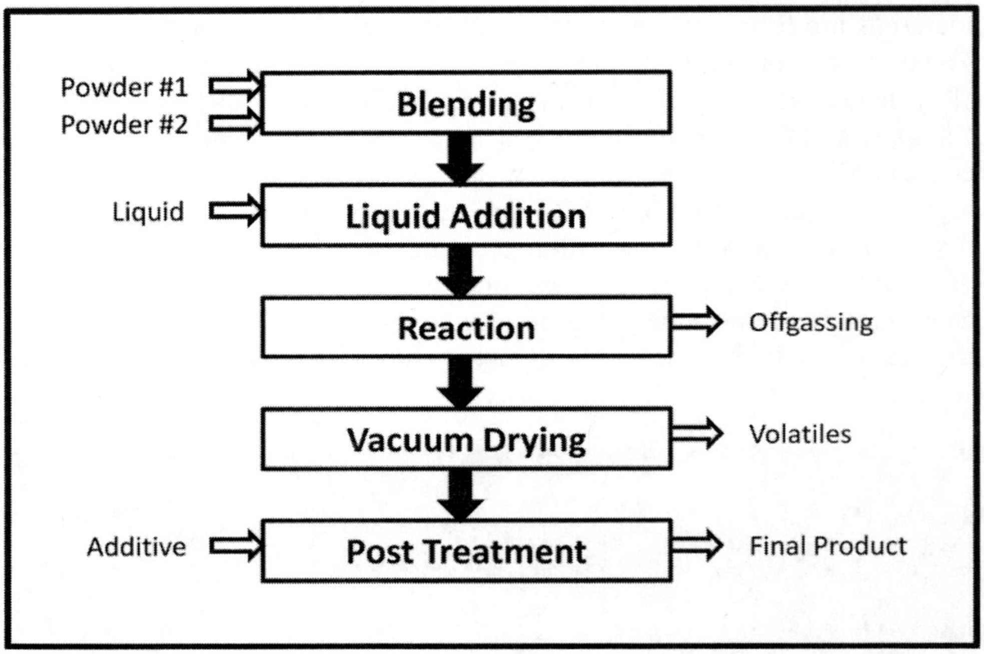

Figure 2: Typical Process Flow

Polymer Spray-Drying

A methacrylate copolymer dissolved in a flammable solvent was pre-heated and sprayed into the plow mixer, which was heated and under vacuum. The flashing solvent was condensed and recovered via the vacuum system, while the polymer formed a fine powder. Once formed, the powder remained fluidized by the action of the plows. This powder, heated by the walls, served as a substrate for additional sprayed product. Particle size was controlled by use of the chopper to break down any agglomerates that formed.

Surface Treatment of Clay

Certain minerals are found to have increased utility when surface-treated with chemical modifiers, such as a quaternary ammonium salt. Clays present a particular problem, in that they tend to become extremely cohesive when wetted. This results in more shear-induced heating, which further increases the cohesion.

In this example, a powdered clay material was first pH-modified with caustic and water, and heated to the desired temperature for reaction. The mixture was agitated at low speed. The quaternary ammonium compound was then metered in over several minutes, while the power draw of the main agitator was monitored. After observing the amperage peak, the agitator speed was increased and the choppers used, while applying vacuum to dry and granulate the product.

Chemical Vapor Deposition

Chemical vapor deposition (CVD) was accomplished, using the mixer as a mechanically fluidized bed. The substrate was first pre-heated and dried under high vacuum. With the substrate fully fluidized, vacuum was used to introduce the volatile reactants into the vapor space, alternating from one to the other in order to build up the desired surface layer.

Direct-Contact CO_2 Chilling

A food customer required rapid blending of dry ingredients with a low-melting oil, resulting in a free-flowing powder. The dry ingredients were pre-chilled with the direct injection of solid CO_2 powder via a "snow horn." With full fluidization, the warmed oil was sprayed in, followed immediately with additional solid CO_2, which quickly cooled the material to the desired packaging temperature.

Discussion

As can be seen by the above case studies, the B&P Littleford plow mixer is a useful and highly configurable processing unit. It is ideal for a variety of unit operations, and at least capable in many others. As such, it can be used to develop "one-pot" processes for the economic scale-up of new batch processes. One multi-purpose piece of equipment will require less capital than several one-use machines, which may or may not be repurposed.

At the Market Development stage, an appropriately sized plow mixer may be tasked to perform the entire process initially. It can also be used to produce sufficient intermediates for use in other equipment testing, for example if a traditional spray dryer might be more economical for the drying step, or a ribbon blender for a final additive and packaging hopper.

From a Process Safety Management perspective, the fewer product transfers done with (potentially) hazardous new products and mixtures, the better. This minimizes the potential for personnel exposure as well as flammability concerns, environmental release, etc. This is true at every stage of process development, but especially in the pilot plant stage, when the volume of chemicals becomes significant, but experience with the process is still relatively low.

481219 Thermodynamic Modeling of Aqueous Nickel Nitrate-Nickel Chloride-Nickel Sulfate System with Electrolytes NRTL Model

Abstract:

Thermodynamic Modeling of Aqueous Nickel Nitrate-Nickel Chloride-Nickel Sulfate System with Electrolytes NRTL Model

Yuan Li, Benjamin Caudle, Chau-Chyun Chen

Texas Tech University

Abstract

Millions of tons of electronic waste such as computers and mobile phones, known as e-waste, are produced each year. Traditionally most e-waste has been disposed of in landfills [1], but in recent years recycling of e-waste is being taken seriously due to environmental concerns and potential financial gains from recovery of materials. In one mode of recycling, metals in e-waste are dissolved in strong acids and then the valuable metals are recovered through electrochemical methods. Nickel is one of the significant metals in e-waste and its dissolution and saturation can significantly affect the efficiency of the recovery system. Therefore, a comprehensive and validated thermodynamic model for nickel salts in strong acids is a necessity for process simulation of e-waste recycling processes.

With a wide applicability from dilute to saturated electrolyte solutions, the electrolytes NRTL (eNRTL) model is one of the most accurate thermodynamic models for aqueous and mixed solvent electrolytes [2-4]. Moreover, the eNRTL model has been successfully validated for various strong acids and mixed acid systems [5-7]. In this work, we extend the eNRTL model and develop a comprehensive thermodynamic model for nickel salts in strong acids including nitric acid, hydrochloric acid, and sulfuric acid.

We first identify the eNRTL binary interaction parameters for aqueous nickel nitrate, nickel chloride, and nickel sulfate electrolyte systems from available literature data using phase equilibrium properties such as vapor pressure, osmotic coefficient, mean ionic activity coefficient, and calorimetric properties such as molar heat capacity and excess enthalpy. Salt solubility data are then used to identify thermodynamic properties for corresponding salt crystals. The binary system models are then integrated with the strong acid models developed with eNRTL [5-7] and further validated against available data for ternary nickel salt - acid – water systems. These ternary system models will form a sound thermodynamic foundation for use in process simulation of e-waste recycle processes.

[1] Robinson, B.H., E-waste: An assessment of global production and environmental impacts. Science of the Total Environment 408 (2009) 183-191.

[2] Chen, C.-C. *et al.*, Local Composition Model for Excess Gibbs Energy of Electrolyte System. Part I: Single Solvent, Single Completely Dissociated Electrolyte Systems. AIChE Journal 28 (1982) 588-596.

[3] Chen, C.-C. *et al.*, A Local Composition Model for the Excess Gibbs Energy of Aqueous Electrolyte Systems. AIChE Journal 32 (1986) 444-454.

[4] Song, Y. and Chen, C.-C., Symmetric Electrolyte Nonrandom Two-Liquid Activity Coefficient Model. Ind. Eng. Chem. Res. 48 (2009) 7788-7797.

[5] Wang, M. *et al.*, Thermodynamic Representation of Aqueous Sodium nitrate and Nitric Acid Solution with Electrolyte NRTL Model. Fluid Phase Equilibria. 407 (2016) 105-116

[6] Hassanjani, S. et al., Comprehensive Thermodynamic Modeling of Complex Mixed-Solvent Electrolyte Systems: An Investigation on Water-Hydrogen Chloride-Methanol Ternary System. In preparation.

[7] Que, H. *et al.*, Thermodynamic Modeling of the Sulfuric Acid - Water - Sulfur Trioxide System with the Symmetric Electrolyte NRTL Model. J. Chem. Eng. Data. 56 (2011) 963-977.

481220 Thermodynamic Modeling of Ion Adsorption in Capacitive Deionization Cell Units with Enrtl Model

Abstract:

Capacitive deionization (CDI) is a technology to remove salts from low salinity brine solution with electrical cell units. The two electrodes in each cell unit adsorb opposite charged ions in the brine water when charged and release them when discharged. Modified Donnan's theory has been used to model the electrical double layer surrounding electrodes. However, non-ideality of ions adsorbed to the electrodes has yet to be adequately accounted for. [1]

Our previous work established a thermodynamic model for polyelectrolyte solutions by extending Manning's limiting law with eNRTL model representing short range interactions. For a polyion that has a charge density larger than a certain critical value, the counterions in the polyion solution are condensed on the surface of the polyion to reduce the charge density. Manning first described the limiting law for calculating counterion activity coefficients in polyelectrolyte solutions [2]. Our model extended the concentration range of salts so that it offers a comprehensive thermodynamic formulation to correlate and extrapolate thermodynamic behavior of all aqueous polyelectrolyte systems, including ion exchange membranes or resins.

The CDI units at equilibrium have the electrodes with counterions electrically attached to the surfaces. The scenario is very similar to the counterion condensation process of polyions. We apply the extended Manning's model to explain the amount of adsorbed salt in a CDI unit. We consider the electrodes as polyions. The polyions and adsorbed ions form a separate phase separated from the bulk brine solution. This polyions phase is then treated as the polyelectrolyte-water binary.

The activity coefficient of counterions on electrodes is calculated with the extended Manning model. From the equality of chemical potential of solute NaCl in the bulk solution and in the electrode polyion phase, we relate concentrations of bulk solution, amount of adsorbed salt, and charge density of the electrodes. Both the sodium cation and chloride anion are considered as counterions because the two electrodes attract oppositely charged ions at the two ends.

Predictions of salt concentrations in the electrodes at different bulk solution concentrations are generated and compared with experimental data. Future works include investigating the relationship between the model parameters and the cell voltage and finding optimal working conditions for a given desired separation.

Reference

1. P. M. Biesheuvel, et al., "Attractive forces in microporous carbon electrodes for capacitive deionization." *Journal of solid state electrochemistry* 18.5 (2014): 1365-1376.
2. Gerald S. Manning, "Limiting laws and counterion condensation in polyelectrolyte solutions I. Colligative properties." *The journal of chemical Physics* 51.3 (1969): 924-933.

481456 Thermodynamic Modeling of the Aqueous Copper(II) - Nitrate - Chloride - Sulfate System with the Electrolyte-NRTL Model

Abstract:

Thermodynamic Modeling of the Aqueous Copper(II) - Nitrate - Chloride - Sulfate System with the Electrolyte-NRTL Model

Benjamin Caudle, Yuan Li, Chau-Chyun Chen

Texas Tech University

Abstract

Electronic waste (e-waste) is waste associated with computers and cell phones, as well as increasingly digitized appliances such as refrigerators and ovens. As demand grows in emerging markets, the need for proper handling of this waste increases as well [1]. A unique characteristic of e-waste is the presence of both valuable materials, such as copper and members of the platinum group, and hazardous materials. Proper recycling of e-waste is necessary to prevent the release of toxic substances into the environment and has the added benefit of yielding materials that can be reused or sold. The most promising method of recycling is through leaching with mixed acids, in which metals can be selectively dissolved depending on the composition of the leaching solution. Copper is typically the most prevalent metal in e-waste and, even if it is not recovered for its own sake, its behavior in solution has a large impact on the system as a whole. For this purpose, a thermodynamic model of copper in common acid solutions is a useful tool for the design of recycling systems.

The electrolyte non-random two-liquid (eNRTL) model has proven to work for a wide range of electrolytes in both single and mixed solvents [2-4]. The nitric, chloric, and sulfuric acid systems already have validated eNRTL models available [5-7], and the prime advantage of the NRTL model is straightforward and accurate extension to other multicomponent systems.

This work extends the eNRTL model to cover the behavior of copper(II) in nitric, chloric, and sulfuric acids and mixtures thereof. Beginning with the binary systems, parameters are regressed from the available literature data. This includes phase equilibrium and calorimetric properties as well as solubility data for the respective copper salts. The binary models are then integrated with the existing acid system models and validated against literature data for the ternary and quaternary systems. The resulting model is an integral part of the overall model for e-waste recycling processes.

[1] Widmer, Rolf *et al.*, Global perspectives on e-waste, Environmental Impact Assessment Review, Volume 25, Issue 5, July 2005, 436-458

[2] Chen, C.-C. *et al.*, Local Composition Model for Excess Gibbs Energy of Electrolyte System. Part I: Single Solvent, Single Completely Dissociated Electrolyte Systems. AIChE Journal 28 (1982) 588-596.

[3] Chen, C.-C. *et al.*, A Local Composition Model for the Excess Gibbs Energy of Aqueous Electrolyte Systems. AIChE Journal 32 (1986) 444-454.

[4] Song, Y. and Chen, C.-C., Symmetric Electrolyte Nonrandom Two-Liquid Activity Coefficient Model. Ind. Eng. Chem. Res. 48 (2009) 7788-7797.

[5] Wang, M. *et al.*, Thermodynamic Representation of Aqueous Sodium nitrate and Nitric Acid Solution with Electrolyte NRTL Model. Fluid Phase Equilibria. 407 (2016) 105-116

[6] Hassanjani, S. et al., Comprehensive Thermodynamic Modeling of Complex Mixed-Solvent Electrolyte Systems: An Investigation on Water-Hydrogen Chloride-Methanol Ternary System. In preparation.

[7] Que, H. *et al.*, Thermodynamic Modeling of the Sulfuric Acid - Water - Sulfur Trioxide System with the Symmetric Electrolyte NRTL Model. J. Chem. Eng. Data. 56 (2011) 963-977.

479968 Careers in Process Development: How Aiche Can Help!

Abstract:

Process development is a career path with high visibility in industry, connecting idea generation through development to commercial deployment. The process development lead must act as an 'intrapreneur" to address all aspects of technology development, from process technology, to footprint and sustainability, and including process economics. This poster will provide an overview of Process Development activities, and how AIChE involvement can provide contacts and skills needed to excel. Also, how can AIChE better serve your needs?

480257 Optimal Design of Mixed Refrigerant System in Ethylene Plants for Maximum Energy Savings

Abstract:

Refrigeration system is one of the most important and critical operating system in the chemical and petrochemical industries. The tradition cascade refrigeration system (CRS) has been used for decades in ethylene plants. However, the use of the multiple refrigerants in the CRS requires a very number of separate compressors, heat exchangers, pumps and associated pipes for each temperature and pressure level. Mixed refrigerant system (MRS) has been introduced for decades, where the pure working fluids are replaced by mixed working fluids. MRS is more efficient in energy saving and less capital cost than CRS and has been widely applied in liquefaction of natural. However, because of the complexities of ethylene manufacturing process and refrigerant systems aforementioned as well as possible intellectual copyright concerns, the in-depth studies for a complex mixed refrigeration system design and operation are still lack of reporting. In this study, a general methodology has been developed for synthesis of MRS with a tertiary refrigerant, a mixture of methane, ethylene and propylene, to replace the ethylene-propylene cycles of the CRS in an ethylene plant. First, the exergy-temperature chart combined with the exergy analysis is presented to comprehensively analyze the thermodynamic nature of traditional CRS to help to locate and evaluate exergy loss in the system. Opportunities to improve are explored and identified in the exergy-temperature chart. Then, a superstructure is built from the conclusions of thermodynamic analysis to cover all the possible arrangements for the cooling supplies. Moreover, a simulation-assisted mixed-integer linear programming (MILP) model is developed and solved for the optimal process synthesis and heat exchange network (HEN) design. The case study about a mixed refrigerant system used in a typical ethylene plant is employed to demonstrate the efficacy of the developed methodology. The shaft work of compressor has been reduced tremendously. And the capital cost and utilities cost have been saved by 32% and 16%, respectively.

480398 Hydrate Formation Experiments in a Lab-Scale Vertical Flowloop

Abstract:

The formation of gas hydrates in subsea oil and gas flowlines remains a significant operational risk, due to their rapid growth rates and ability to form large aggregates. Much of our understanding of hydrate blockage formation to date has been derived for oil-continuous pipes, where water droplets may be dispersed in the oil phase. However, knowledge of hydrate growth rates and particle transportability in water-continuous systems is less well-understood. In this presentation, we employ the use of a vertical flowloop to investigate hydrate growth behavior using mineral oil and water under steady-state and transient operations.

A high-pressure water tunnel (HPWT), initially constructed to examine hydrate growth behavior on rising gas bubbles following a subsea blowout, was converted for use as a vertical flowloop. Changes included the installation of an ultrasonic flowmeter, differential pressure transducer and viscometer. More recently, the flowloop has been adapted to allow the introduction of mineral oil, as well as the installation of a high-pressure wire feedthrough to measure the conductivity of the flowing fluids, thereby determining the continuous phase.

Experiments were performed using water and mineral oil at water volume fractions of 100, 80, and 50%, at a constant mixture velocity of approximately 1 m/s. Gas addition, hydrate dissociation and flowloop restart tests (with and without hydrates present) were conducted to investigate the effects of hydrate formation under both steady-state and transient operations. A transparent 'bulls-eye' pipe segment, on one of the vertical sections of the loop, allowed for visual observation of hydrate growth behavior. Coupled with a high-speed camera, this allowed for detailed investigation of interfacial phenomena under flow. Contrary to the formation of solid shells on water droplets in oil-continuous systems, we present new evidence suggesting hydrates do not always form complete shells around methane-saturated oil droplets in water-continuous systems. Instead, hydrate crystals were sheared from the gas-water interface soon after formation, and aggregated in the continuous water phase.

These tests have enabled a quantitative comparison of the potential mass transport limitations to hydrate growth, including gas dissolution in the aqueous phase and gas diffusion to the growing hydrate particle. The measured hydrate slurry viscosity increased directly with hydrate volume fraction, increasing the momentum energy required to restart flow following a shut-in. The results from this investigation provide new insight into hydrate growth behavior in high watercut systems, which may be used in the validation of predictive models.

480410 Reduction of Hydrate Film Growth Rate and Cohesive Force By Surface Active Components in Crude Oil

Abstract:

During the production of offshore oil and gas, the cooling of hydrocarbons toward seafloor temperatures enables the formation of gas hydrates, which may restrict fluid flow and ultimately block the flowline. As hydrates aggregate and begin to form a blockage, a hydrate film may grow between individual particles or between hydrate particles and the pipeline wall, respectively resulting in a higher slurry viscosity or a reduced hydraulic diameter. The hydrate cohesive force and hydrate film growth rate are fundamentally controlled by the properties of the interface. While it has been shown previously that hydrate cohesive force may be affected by naturally-occurring surfactants in the crude oil, there is limited data available to inform whether hydrate film growth rate may also be affected by these components. In this study, we have used a micromechanical force (MMF) apparatus to quantify the film growth rate of cyclopentane hydrate at a moderate subcooling. Naturally-occurring surface active species were obtained from two Australian crude oils by solvent extraction, resulting in the separation of three oil fractions: asphaltenes, binding resins, and free resins. Each oil fraction was then added to a chemically inert hydrocarbon phase, and was injected in the MMF experimental cell to quantify its impact on hydrate film growth rate. At mass fractions below 500 ppm, the hydrate film growth rate was reduced by at least one order of magnitude by asphaltenes; both resin classes were less effective at reducing hydrate film growth rate. Despite variations in the film growth measurements, cohesive force trials with the MMF revealed that each of the three fractions was able to reduce the hydrate cohesive force by two orders of magnitude. The presence of each fraction also caused an increase in the wetting angle of the water droplet on the hydrate particle surface, which suggested that these naturally-occurring components may adsorb to the hydrate particle surface.

480411 A New Transient Framework for Assessing Hydrate Formation

Abstract:

As the production of oil and gas moves toward deeper water, the problem of hydrate formation in flowlines becomes more prominent. Traditionally, a strategy of avoidance requiring the addition of large volumes of thermodynamic inhibitors has been pursued. The fiscal constraints of deepwater production may be overcome if hydrates are allowed to form, and the risk of blockage can be successfully managed. This evolving view posits that the flowline may be allowed to operate within the hydrate equilibrium region, while employing both unique operating strategies and novel technologies to avoid blockages. For such methods to be applied with confidence, a robust understanding of both hydrate formation and multiphase transport behaviour are required. The current engineering approach models both concepts independently, where the effect of hydrate growth on slurry transportability is typically isolated from transient flow simulations. In this work, we introduce a new approach to bridge this gap by incorporating the effect of pressure loss due to slurry viscosification and stenosis, or a narrowing of the flow channel, on the driving force for multiphase flow. We present the results of a deepwater flowline case study, in which established hydrate formation and transportability calculations are coupled to a fully transient multiphase flow engine; this unique approach enables flow assurance engineers to evaluate the severity of hydrate formation on a physical basis. For the first time, the quantitative effects of operating within the hydrate stability region can be estimated with respect to hydrate growth, production rate, and fluid properties.

480413 Rheological Characterisation of Hydrate-in-Oil Slurries

Abstract:

Gas hydrates are ice-like solids that may form in crude oil flowlines under high pressure and at low temperature, resulting in high frictional pressure loss and increasing the risk of a non-flowing condition. The viscosity of hydrate slurries has been suggested as a primary criterion to determine blockage risk, but limited rheology data are available to inform the predictive viscosity models. The rheology of hydrate slurries formed from water-in-oil emulsions was measured using a high-pressure controlled-stress rheometer, where gas saturation in the slurry as maintained with a vane-blade rotor. Tests were conducted under isochoric conditions, where the apparent viscosity and pressure of the system were recorded throughout the hydrate growth process, enabling the first comparison of hydrate slurry viscosity as a function of hydrate volume fraction in the liquid phase. The results show that, for 5 to 30 % watercut systems, hydrate formation was repeatedly identified by a significant decrease in pressure and a corresponding increase in slurry viscosity; the maximum and steady-state viscosity values recorded during the hydrate formation process deviated by a maximum of 10% through independent trials. As vane blade rotor was able to maintain saturation of the crude oil phase, these measurements represent the first insight into the worst-case slurry viscosity condition achievable in crude oil flowlines.

Modelling Salts Effect on Hydrate Inhibition with CPA-Electrolyte and Pseudo-Salt Approach with Excess Gibbs Energy Mixing Rules

Xiaohong Zhang, Nuno Pedrosa, Richard Szczepanski
KBC Advanced Technologies Ltd (A Yokogawa Company), London, UK

Abstract

Gas hydrate formation analysis and management have become an essential part of engineering work for the correct design, commissioning and operation of offshore and onshore production and transportation.

The conventional engineering approach for suppression of hydrates formation is to add inhibitors such as methanol or glycol to alter the thermodynamic equilibrium so that hydrates cannot form under production conditions. Although there has been much work recently on low-dosage kinetic inhibitors, there are many situations where it remains essential to use thermodynamic inhibitors. In order to accurately predict the hydrate phase behaviour and the dosage of hydrate inhibitors, the equation of state used has to be able to model systems where hydrocarbons, water and other polar components such as methanol are involved, as well as be able take into account of the presence of salts, and their effect on the solubility of water and hydrate inhibitors, and the accumulated hydrate inhibition.

Traditionally cubic equations of state, such as Peng–Robinson (PR) or Redlich–Kwong–Soave (RKS), with simple mixing rules, are not able to predict correct phase behaviour of highly polar mixtures like water and salts. In order to model not only gas and oil phase but must also an aqueous phase with mixed solvents of alcohols, glycols and salts, either complex mixing rules based on Excess Gibbs energy or additional association terms are needed to provide reliable prediction on complex phase equilibria between the fluid and solid phases. In this study two well established models; CPA (Cubic-Plus-Association) with Electrolyte and the advanced version of RKS with modified NRTL type mixing rule are applied to model the effect of salts and the mixed-solvent electrolytes involving methanol on gas hydrate inhibition.

The study will compare the performance and predictability of the two well-developed models, addressing the importance of accurate fluid model for the prediction of hydrate formation and highlight the best practice on modelling the accumulated effect of salts on hydrate inhibition.

1 Introduction

Gas hydrates are solid ice-like compounds formed by water and stabilised by the presence of "guest" molecules within the hydrate lattice cavities [1]. As these guest molecules are typically light gases and hydrocarbons, hydrates are of particular importance to the oil and gas industry. Although hydrates may be an important future hydrocarbon energy resource with the development of appropriate means of capture, transportation and storage, their presence in oil and gas exploration and production is a major challenge due to the operational and safety concerns. Hydrates can exist in-situ in petroleum reservoirs where they can cause blowouts if drilled into inadvertently[2]. Alternatively, extreme conditions of temperature and pressure (typically low temperatures and high pressure) mean that hydrates can easily form during the drilling process if fluids containing water, drilling muds, seawater or produced water, come into contact with reservoir fluids[3,4]. Formation of solid hydrates can also cause blockages in subsea risers, blow out preventers (BOPs), choke and kill flowing lines. Conditions during well shut-in are particular in favour of hydrate formation if high pressures are combined with falling temperatures[1].

To prevent and suppress the gas hydrate formation in oil and gas exploration and production, the conventional engineering approach is to use hydrate inhibitors such as methanol or glycols (MEG, DEG, TEG) to alter the thermodynamic properties of the aqueous phase so that hydrates cannot form under the production conditions[5]. During the production, the effect of the presence of salts in produced or sea water must also be considered. This work will show that their effect cannot be neglected. Use of thermodynamic inhibitors, however, is expensive and can lead to high capital expenditure for storage and significant operational costs. Thus it is desirable to understand the mechanism of thermodynamic hydrate inhibition so that correct operation schemes to avoid over-dosage can be implemented to control and prevent hydrate formation economically.

To have a correct operation scheme and avoid over-dosage, thermodynamic modelling is essential for helping understanding the phase behaviour of the fluid and the inhibitors, and most importantly the effect of the additional salts on hydrate inhibition as these ionic salts can act as hydrate inhibitors and are always present in seawater, formation and produced water under the drilling and production conditions. In this work, two well-developed thermodynamic models in Multiflash^{TM} are applied to model the effect of ionic salts on the hydrate inhibition; the CPA (Cubic-Plus-Association) with Electrolyte and the advanced version of RKS with NRTL type mixing rule (RKSA-Infochem)[6,7].

2 Modelling Gas Hydrates and Hydrate Inhibition

2.1 Gas Hydrates Structures

The driving force of gas hydrate formation depends on the three key criteria; the presence of water, hydrate forming components and the right conditions of pressure and temperature. Gas hydrates form when the criteria are satisfied and water forms a cage-like structure around a "guest" molecule. There are many compounds that can form hydrates and several hydrate structures. However, in the oil and gas industry only three forms of hydrate structure need to be considered; Structures sI, sII and sH[1]. Among those, the structure I hydrate is the simplest structure with two types of cavities, the components that are found to form structure I hydrates in oil and gas production and processing are methane, CO_2, H_2S and ethane. Similar to structure I hydrate, the structure II hydrate also has two types of cavities but is more complex and much larger, therefore large molecules such as propane and iso-butane can fit into the cavities of the hydrate structures to form structure II hydrates[8,9]. The structure sH is the last hydrate structure that is relevant to the oil and gas industry. Unlike Structures I and II, structure H is never formed by a single pure component but requires two hydrate formers, a small molecule such as methane and a larger Type H forming molecule. Typical of the latter are branched butanes and cyclics. Structure H is made up from three types of cage: a regular dodecahedron, an irregular dodecahedron and an irregular 20 sided polyhedron[10].

The larger molecules of a gas mixture normally determine the hydrate structure formed so the presence of propane and i-butane in natural gas mixtures means that the normal form found in gas production and processing is Structure II hydrate (sII).

2.2 Gas Hydrate Forming Components

Gas hydrate formation requires small components to occupy the hydrate lattice to stabilise the hydrate structures. As one knows, many of the components in natural gas, gas condensates or oils can form gas hydrates but not all of them are not explicitly identified in compositional analyses and therefore will not be included in any modelling of hydrate formation. To develop a hydrate model for a typical component, the relevant hydrate dissociation data are required for establishing the model parameters[1]. However, the most common hydrate forming components relevant to oil and gas production and processing are well-understood and can be also modelled by the PVT software, Multiflash^{TM}. The detailed list of the hydrate forming components is summarised in Table 1, where the components marked with star are large molecules, requiring small molecules like methane to help forming hydrate structures.

Natural gases	Condensates/oils	Process
Nitrogen	Benzene*	Ethylene
Carbon dioxide	Cyclopentane*	Propylene
Hydrogen sulphide	Cyclohexane*	Other olefins
Methane	Methylcyclopentane*	
Ethane	Cycloheptane*	
Propane	Methylcyclohexane*	
Butane*	Ethylcyclopentane*	
Isobutane	Cyclooctane*	
Neopentane*	1,1 Dimethylcyclohexane*	
	cis 1,2 Dimethylcyclohexane*	
	2 Methyl butane*	
	2,2 Dimethyl butane*	
	2,3 Dimethyl butane*	
	2,2,3 Trimethyl butane*	
	3,3 Dimethyl pentane*	
	2,2 Dimethyl pentane*	
	2,3 Dimethyl-1-butene*	
	3,3 Dimethyl -1-butene*	
	3,3 Dimethyl-1-butyne*	
	cis cyclooctene*	
	bicyclo[2,2,2] oct-2-ene*	
	Adamantane*	

Table 1 Hydrate forming components. *helper gas needed to form hydrates

2.3 *Gas Hydrates and Hydrate Inhibition Calculation*

The calculation of phase equilibria involving hydrates is, in principle, no different from any other phase equilibrium calculation. The phase equilibrium equations must be solved to satisfy the equality of fugacities of each component in every phase:

$$f_{i,1} = f_{i,2} = f_{i,3} = \ldots = f_{i,p} \tag{1}$$

For a multi-component system, solving the equation (1) is accurately very challenging and requires a robust and complex phase equilibrium algorithm involving not only gas hydrate phases but also other fluid and solid phases[6]. Usually it is necessary to consider up to six phases: gas, hydrocarbon liquid, aqueous liquid, ice, Hydrate I and Hydrate II. To model the effect of salts on hydrate inhibition, additional solid salt phases are also required, leading up to total of 17 phases that may be considered. The types of models typically used to describe the properties of each of these phases in Multiflash™ are briefly described in the following sections.

2.3.1 *Modelling Gas Hydrates*

The gas hydrate model developed in Multiflash™ is based on the work of Van der Waals and Platteeuw[11,12]. It is assumed that the formation of gas hydrate is thermodynamically as having two steps: (1), pure liquid water solidifies to form an empty hydrate lattice; (2), gas molecules are adsorbed into the cavities in the hydrate lattice. The former is analogous to the formation of ice and is described by the same thermodynamic equations, although the parameter values of hydrate lattice are different. The latter is described by the statistical mechanics of the hydrate adsorption process that was derived by Van der Waals and Platteeuw for modelling the interaction between the gas molecules and the hydrate lattice using an effective intermolecular potential. The general mathematic formulation for describing the change of the difference in chemical potential between an empty and filled hydrate lattice is given by[1,6]

$$\frac{\Delta\mu^H}{RT} = \sum_k v_k ln\left(1 - \sum_i n_{i,k}\right) \tag{2}$$

where the first summation is over the number of cavities k in hydrate lattices, and the second one is the summation over the component, *i*. The $n_{i,k}$ is the probability that a cavity, *k*, is occupied by a gas or light hydrocarbon molecule, *i*, which is defined by

$$n_{i,k} = \frac{C_{i,k} f_i}{1 + \sum_j C_{j,k} f_j} \tag{3}$$

The f_i is the fugacity of the component, *i*, that is calculated by thermodynamic fluid models such as CPA and RKSA-Infochem. $C_{i,k}$ is the Langmuir constant of the component, *i*, which is cavity specific, temperature-dependent property.

$$C_{i,k} = C_{i,k}(T, \boldsymbol{a}_{i,k}) \tag{4}$$

where the adjustable parameters $a_{i,k}$ are estimated by matching experimental hydrate dissociation data. This approach has been widely used and tested for many years and is capable of accurately representing the experimental data on hydrates. The gas hydrate model in MultiflashTM has been developed with large sets of experimental data including proprietary data from industrial gas hydrate projects and the data from public domain[1]. Later the model was extended and further improved as part of a deepwater JIP project[13]. In general the model can give accurate prediction of hydrate dissociation temperature within one degree. The examples of the performance of the models are shown in Figure 1 and Figure 2.

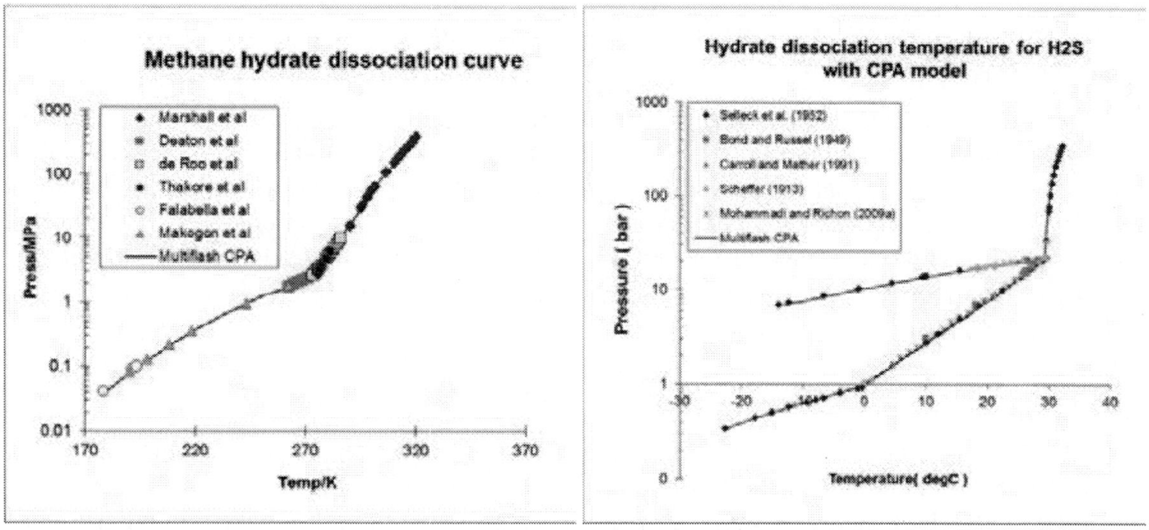

Figure 1 Examples of the Gas Hydrate dissociation curves of Methane and H_2S predicted by Multiflash™[1]

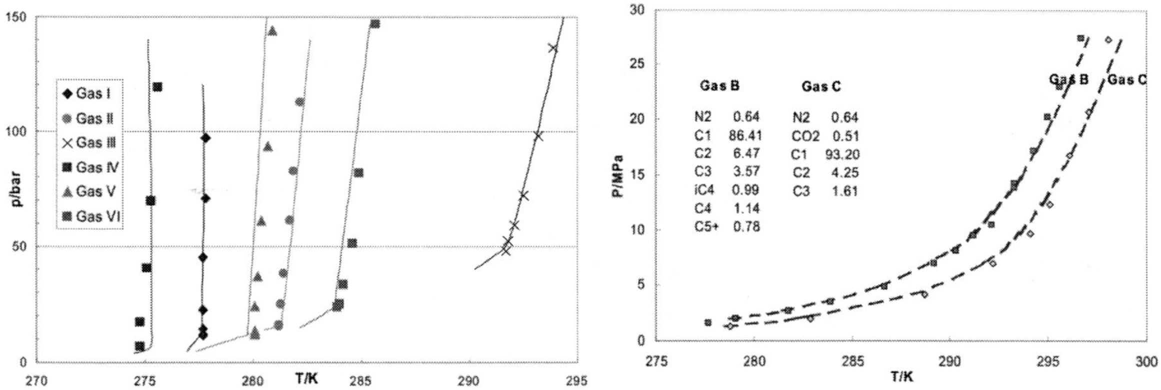

Figure 2 Examples of the Gas Hydrate dissociation curves of natural gases and condensates predicted by Multiflash™[1]

2.3.2 Fluid Phases

As described in equations (2) and (3), the fluid models play an important role in gas hydrate calculations where thermodynamic fluid models provide the fugacity calculations of the hydrate forming components as well as for the fluid phases[1,6,14]. Thus modelling the gas hydrate phase behaviour accurately demands accurate modelling for not only gas and oil phase, but must also the aqueous phase that will usually contain alcohols, glycols, electrolytes and several possible solid phases.

The general fluid modelling approach in Multiflash™ is to enhance the basic cubic equation of state (EOS) by adding appropriate thermodynamic terms to the total residual Helmholtz free energy. Hence the general fluid phase model in Multiflash™ has the following structure

$$A^{res} = A^{res}_{RKS} + A_{assoc} + A_{electrolyte} \tag{5}$$

where A^{res} is the total residual Helmholtz energy of the fluid phases, A_{assoc} is the contribution from chemical association and $A_{electrolyte}$ is the term that accounts for contributions of electrolyte solutions. A^{res}_{RKS} is the residual term described by the cubic EoS, e.g. Redlich-Kwong-Soave EoS (RKS).

The aqueous phase is perhaps the most difficult to represent, particularly if inhibitors such as methanol and/or salts are present. Having a reliable model to accurately describe the aqueous phase behaviour is crucial for modelling gas hydrates. However, the aqueous phase cannot be adequately modelled with a conventional cubic EOS. Multiflash^{TM} offers the advanced version of RKS (RKSA) model either with an NRTL type mixing rule (RKSA-Infochem) or with an explicit association term (CPA)[14]. The common cubic EoS A^{res}_{RKS} term based on RKS can be expressed by

$$A^{res}_{RKS} = -NRTln\left(1 - \frac{b}{V}\right) - N\frac{a}{b}ln\left(1 + \frac{b}{V}\right) \tag{6}$$

For the RKSA-Infochem approach, only the first term in equation (5) is considered that includes the "a" parameters determined by fitting the vapour pressure to improve the saturation pressure prediction and the Peneloux volume correction is also added to improve the liquid density[7]. The modified NRTL-Infochem mixing rule then enables the RKSA to model the phase behaviour of the polar components such as water, methanol and glycols. The form of the modified NRTL Infochem mixing rule is:

$$a = b\sum_i n_i \frac{\sum_j n_j b_j \xi_{ji} G_{ji}}{\sum_j n_j b_j G_{ji}} \tag{7}$$

Where

$$\xi_{ji} = \frac{2\sqrt{a_i a_j}(1 - k_{ji})}{b_i + b_j} \tag{8}$$

And

$$G_{ji} = exp\left(\frac{\alpha_{ji}(\xi_{ji} - \xi_{ii})}{RT}\right) \tag{9}$$

In general $k_{ji} \neq k_{ij}$ for highly polar components and $\alpha_{ji} = \alpha_{ij}$. The advantage of this approach is that if α_{ij} is set to zero, the mixing rule reduces to the conventional Van der Waals mixing rule that is normally used with the standard cubic EoS.

$$a = \sum_{ij} n_i n_j \sqrt{a_i a_j}(1 - k_{ji}) \tag{10}$$

Where the binary interaction parameters are now symmetric, $k_{ji} = k_{ij}$.

With the binary interaction parameters (BIPs), this NRTL-Infochem mixing rule enables the cubic EoS to handle the properties of hydrate inhibitions in a thermodynamically consistent way allowing all the effects of adding inhibitors to be modelled including the depression of hydrate points, the depression of the ice formation and the reduction in the vapour pressure of water.

In Multiflash^{TM}, the chemical association approach (CPA) based on the Advanced version of RKS (RKSA) is the recommended model for gas hydrate formation and inhibition modelling. The general numerical formulation of CPA in Multiflash^{TM} is based on the work of Kontogeorgis et al [14], which has the form:

$$A_{assoc} = \sum_i n_i \sum_{A_i} \left[lnX_{A_i} - \frac{1}{2} X_{A_i} + \frac{1}{2} \right] \tag{11}$$

Where α_{ij} is the mole number of component, i, and X_{A_i} is the mole fraction of molecules not bonded at the association site A_i, given by

$$\frac{1}{X_{A_i}} = 1 + \rho \sum_j n_j \sum_{B_j} X_{B_j} \Delta_{A_i B_j} \tag{12}$$

Where $\Delta_{A_i B_j}$, the association strength is expressed as

$$\Delta_{A_i B_j} = b_{ij} \beta_{A_i B_j} g(\rho) \left[exp \left(\frac{\varepsilon_{A_i B_j}}{RT} \right) - 1 \right] \tag{13}$$

Where $g(\rho)$ is the radial distribution function. The association energy $\varepsilon_{A_i B_j}$ and the association volume $\beta_{A_i B_j}$ are the two parameters that are determined by fitting the experimental VLE and thermal data. For the details of the CPA model, refer to the work by Kontogeorgis et al [14].

In general both the RKSA-Infochem and CPA models are capable of giving accurate predictions on the fluid and solid phases. But CPA offers better prediction of the methanol partitioning and more reliable extrapolation performance in the region outside of the conditions covered by the experimental data.

2.3.3 Salinity Effect on Gas Hydrates Formation

Salts, as a natural hydrate inhibitor, can be as effective as methanol and glycols on hydrate inhibition by reducing the water activity in aqueous solution to supress the hydrate formation[17-19]. Thus modelling salt inhibition on gas hydrate formation accurately is important for the oil and gas exploration and production as the formation water and drilling muds may contain concentrated solutions of sodium chloride, calcium chloride or other salts.

The electrochemistry of aqueous salt solutions is complex but for the main objective of modelling salts effect on hydrate inhibition, it is reasonable to simplify the complexity, at the same time, and capture the key features of the salts effect on the properties of the fluid and the hydrates phase behaviour. Thus in order to model the salinity effect on hydrate formation accurately, the model must accurately predict: (1): salts stay in aqueous phase; (2), salts precipitate when solubility limit is reached; (3), depression of hydrate and ice formation points; (4), salting-out effect; (5), presence of salts reduces water vapour pressure as well as ensure the approach to be thermodynamically consistent with the gas hydrate model and the fluid models. In Multiflash™, there are two practical approaches available that can be applied to model the effect of salts on hydrate formation; one is the conventional electrolyte approach with CPA and another one is a pseudo equivalent-NaCl salt component approach with RKSA-Infochem[7].

The electrolyte approach with CPA in Multiflash™ is based on the Debye-Hückel theory plus virial corrections and Born term to predict the effect of salts on hydrate inhibition[14-16]. The general electrolytes contribution to the total residual Helmholtz free energy is:

$$A_{electrolyte} = A_{DH} + A_{VR} + A_{Born} \tag{14}$$

Where A_{DH}, the Debye-Hückel describes the electrostatic ion-ion interactions. A_{Born} for the interactions between ions and the surrounding medium, and A_{VR}, the virial terms for the short range interactions between the ions. The Debye-Hückel term, the contribution of the electrostatic interactions between charged ions to the total residual Helmholtz free energy can be written in the following form:

$$A_{DH} = -\frac{4A_0V}{Z^+Z^-}\sum_{i,j} n_i n_j z_i z_j \frac{\tau_{ij}}{\left(b_{ij}^0\right)^3} \tag{15}$$

Where A_0 is a constant, Z^+ and Z^- are the total positive and negative charge respectively. The index i represents the cations, and j the anions, and z_i is the charge of the ions. b_{ij}^0 is specific Debye-Hückel parameter for each pair of ions and τ_{ij} is defined as:

$$\tau_{ij} = ln\left(1 + y_{ij}\right) - y_{ij} + 0.5\, y_{ij}^2 \tag{16}$$

and

$$y_{ij} = \frac{b_{ij}^0}{\sqrt{V\epsilon T}}\Phi^{1/2} \tag{17}$$

where V is the total volume, T is the temperature, ϵ the relative permittivity of the solvent and Φ the ionic strength defined as

$$\Phi = \frac{1}{2}\left(\sum_{i\in cation} n_i z_i^2 + \sum_{j\in anion} n_j z_j^2\right) \tag{18}$$

In this electrolyte approach, it is assumed that the salts are fully dissociated into cation and anion ions in aqueous solution. The presence of the charged ions causes the electrolyte solution to deviate much more from ideal solution behaviour than a non-electrolyte solution does. For simple electrolytes, water is the only solvent and the electrolytes are solute. In order to model the effect of salts on hydrate inhibition in oil and gas production environment, however, the aqueous solution usually contains other species like water, methanol, glycols and salts as well as other dissolved acidic gases. Therefore the interactions between ions and the surrounding solvents have to be described by "mixed solvents" or pseudo solvents, and such interactions can be expressed by Born equation for describing the contribution to the total residual Helmholtz free energy in terms of the presence of polarizable solvents affecting the relative permittivity of the dielectric medium. In the implementation of Multiflash^{TM} CPA-Electrolyte, the second and third virial contributions are also considered. For the details of Born equation and virial corrections, refer to the scientific textbooks[14-16].

With the additional virial corrections and Born term, the model is able to accurately predict the suppressing effect of the mixed solvents (water, methanol and glycols and salts) on hydrate formation, the solubility of gases in salted water, the freezing point depression and vapour pressure lowering, see in Figure 3.

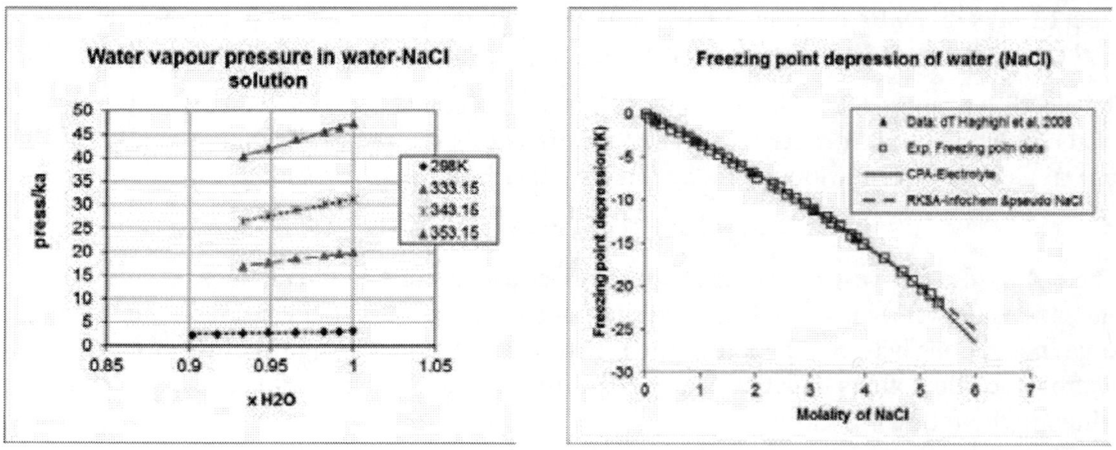

Figure 3 Effect of NaCl on water vapour pressure and freezing temperature, predicted by Multiflash^{TM}[17,18].

On the other hand, the pseudo-salt component approach based on RKSA-Infochem follows the cubic EoS framework by assuming that the salts are represented by a single pseudo NaCl-equivalent salt component (pseudo-NaCl) with physical properties and binary interaction parameters that ensure the model captures the key features of the salts effect on ice, hydrate formation and the properties of fluid phases. The hypothetic pseudo salt component (pseudo-NaCl) behaves like a typical non-volatile and heavy component with high critical temperature and low critical pressure (991.7K,1.86MPa), staying in the aqueous liquid phase or precipitating as a solid when its solubility limit is reached. Setting the interaction parameters of the pseudo-NaCl with hydrocarbons and light gases to a suitable large value also ensures that the salt component, pseudo-NaCl will not enter the hydrocarbon liquid phase.

For salt mixtures containing several types of salts, such as formation water and produced water or brines, rules are required to convert the mixed salt solution into a single equivalent-NaCl solution so that the salinity effect can be modelled by a single pseudo-NaCl[7]. This conversion assumes that the reduction of the water activity by the mixed salts is the same as that by the single pseudo-NaCl in the aqueous solution. Two typical examples of the salt composition of Forties formation water and the North Sea brine are given in Table 2, and the equivalent amount of the pseudo-NaCl for representing the mixed salts is also included in the table.

Salt names	**North Sea Brine, wt%**	**Forties Formation Water, wt%**	**Synthetic Formation Water, wt%**
NaCl	2.354	6.993	12.06
$CaCl_2$	0.116	0.735	1.04
$MgCl_2$	0.524	0.186	0.12
KCl	0.086	0.066	0.56
Na_2SO4	0.428	-	-
$SrCl_2$	-	0.099	0.04
$BaCl_2$	-	0.036	0.01
Equivalent pseudo-NaCl for 100g of water	3.362g	8.637g	15.80g

Table 2 Examples of salts analysis[1]

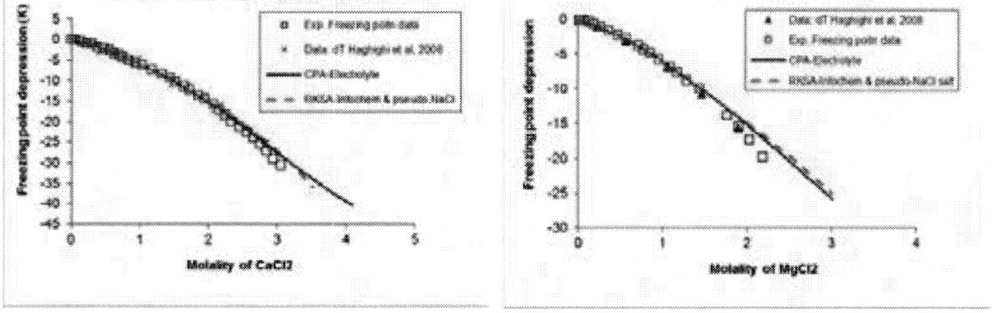

Figure 4 Freezing point depression of water with $CaCl_2$ and $MgCl_2$, predicted by Multiflash^{TM}[18]

The Figure 4 shows the freezing point depression of water as a function of the molality of $CaCl_2$ and $MgCl_2$ respectively, are well predicted by both the CPA-Electrolyte and pseudo-NaCl salt component approach with RKSA-Infochem model.

3 Results and Discussion

In this section, both the CPA-Electrolyte and pseudo-NaCl with RKSA-Infochem model are applied in the investigations of the salinity effect on hydrate formation for several types of mixtures, ranging from simple binary systems to a natural gas, gas condensate and black oil.

Before the calculations, the salinity analysis is necessary to obtain the relevant amounts of ions and the single pseudo-NaCl salt component to the relative amount of water in the mixtures.

3.1 Methane Hydrate Formation in Aqueous Mixed Electrolyte Solutions

The methane structure I hydrate has been investigated for over a century and many experimental measurements have been carried out, covering a wide range of temperature and pressure as shown in Figure 1[1]. To investigate the effect of salts and mixed electrolyte solutions on methane hydrate formation, the data measured by Dholabhai[19] are considered. The results of the hydrate dissociation curves predicted by Multiflash are presented in three groups; NaCl and KCl mixed electrolytes, NaCl and $CaCl_2$ mixed electrolytes, and NaCl, KCl and $CaCl_2$ mixed electrolytes. The results are summarised in Figure 5, Figure 6 and Figure 7 respectively.

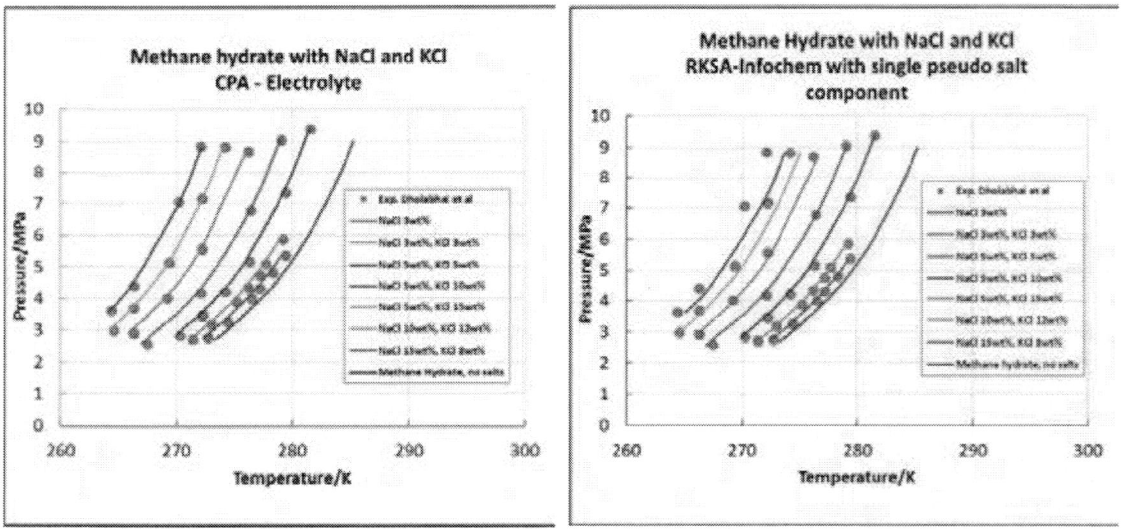

(a) (b)

Figure 5 Methane hydrate dissociation curves with NaCl and KCl. (a), solid lines are predicted by CPA-Electrolyte in Multiflash. (b), solid lines are calculated by RKSA-Infochem with a single pseudo-NaCl salt component.

As clearly shown in Figure 5 (a), the suppression by the mixed electrolytes of NaCl and KCl on hydrate dissociation curves are well predicted by the Multiflash CPA-Electrolyte model, the accuracy of the dissociation temperatures predicted is within 0.5 degree. It is very interesting to see that the sodium chloride by weight is more effective than that of potassium chloride to suppress the hydrate formation. Similar results are also obtained with the single pseudo-NaCl approach, and the predicted hydrate dissociation curves are consistent with the experimental data up to 20wt% of the total salt concentration. However with increasing the salt concentration over 20wt%, the uncertainty of the predicted hydrate dissociation temperatures tends to increase from 0.5 degree up to ~1.5 degree.

For the mixed electrolytes containing NaCl and $CaCl_2$, the predicted results are presented in Figure 6 together with the experimental data. As shown in Figure 6, both models give the consistent and accurate predictions of the effect of salts on hydrate formation as compared with the experimental data. This demonstrates that not only the CPA-Electrolyte model but also the approximated single pseudo-NaCl approach is capable of representing the mixed salts effect on hydrate formation with the total salt concentration up to 21 wt%. The results also show that the effect of the mixed salts with NaCl 6wt% and $CaCl_2$ 10wt% gives the same effect on suppressing hydrate formation as that of NaCl 10wt% and $CaCl_2$ 6wt%.

For the three-salt mixed electrolytes with NaCl, KCl and $CaCl_2$, the CPA-Electrolyte model is slightly better than that of the single pseudo-NaCl approach as shown in Figure 7.

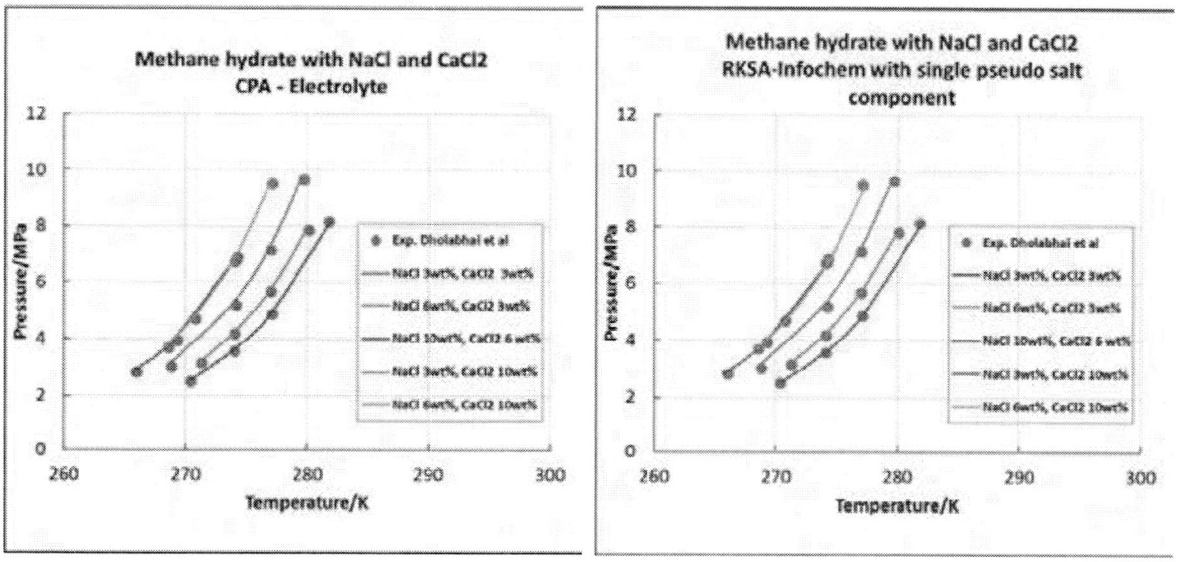

(a) (b)

Figure 6 Methane hydrate dissociation curves with NaCl and $CaCl_2$. (a), solid lines are predicted by CPA-Electrolyte in Multiflash™. (b), solid lines are calculated by RKSA-Infochem with a single pseudo-NaCl salt component.

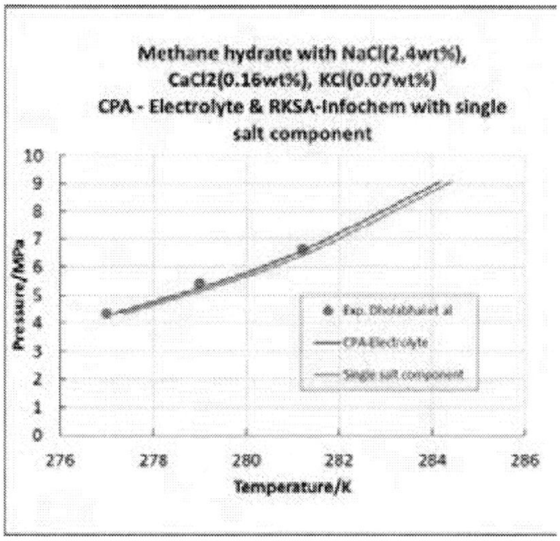

Figure 7 Methane hydrate dissociation curves with NaCl, KCl and $CaCl_2$.

3.2 Methane-CO_2 with Single/Mixed Electrolyte Solutions

A binary methane-CO_2 gas mixture with single and mixed electrolyte solutions is considered to validate the two models. The data of hydrate dissociation curves with single and mixed electrolyte solutions by Dholabhai *et al*[20] were used in the study. The concentrations of methane and CO_2 are fixed to be 80mol% and 20mol% in salt free gas mixture respectively. The single electrolytes are NaCl, KCl and $CaCl_2$ with concentrations ranging from 5wt% to 20wt%, whereas the mixed electrolyte consists of NaCl 5wt%, KCl 5wt% and $CaCl_2$ 10wt%.

The phase behaviour of methane-CO_2-water mixtures is complex involving vapour, hydrocarbon liquid, liquid water, ice, hydrates. Additional solid phases for salts are also needed when the salts are added to the systems. The VLE and the hydrate dissociation curves predicted by Multiflash CPA are presented in Figure 8. As shown in the figures, the phase boundary of both the VLE and hydrate dissociation curves depends on the composition of the methane and CO_2 in the mixture. Both methane and CO_2 forms structure I hydrate but the size, the stoichiometric ratio (number of water molecules to the number of gas molecules) between the methane and CO_2 hydrate is different, leading to different hydrate phase behaviour with temperature and pressure.

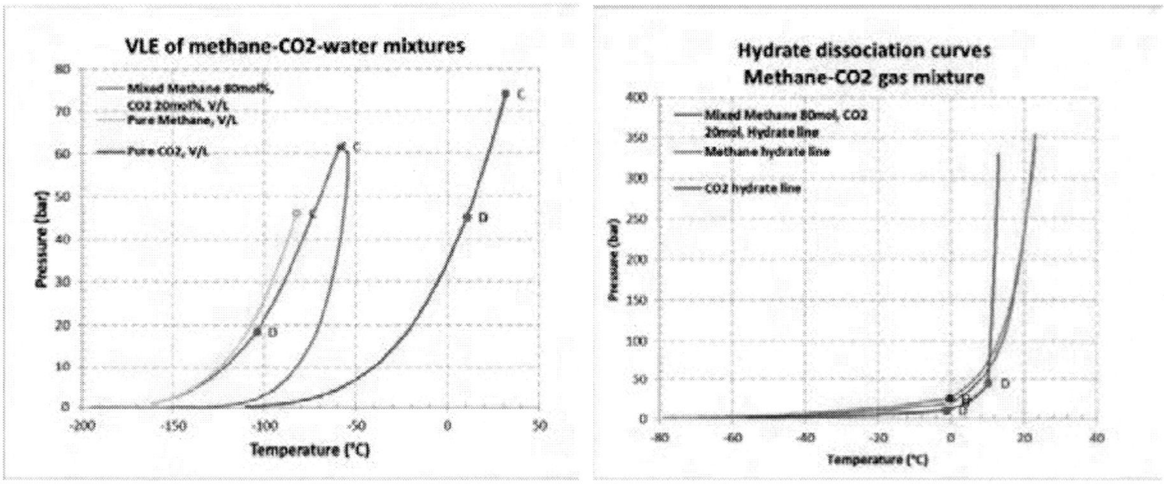

Figure 8 The VLE and hydrate dissociation curves of Methane-CO_2-Water mixtures, predicted by Multiflash CPA.

The hydrate dissociation curves of the methane-CO_2 gas mixtures with salts predicted by Multiflash are summarised in Figure 9 to Figure 12 respectively. For all the cases, the CPA-Electrolyte model gives very accurate predictions of the suppression effect of salts on the hydrate dissociation curves with uncertainty of the dissociation temperature within 0.5 degree.

For the RKSA-Infochem plus a single pseudo-NaCl approach, the hydrate dissociation curves predicted by the model is quite consistent with the experimental data with average uncertainty of dissociation temperature within 1.0 degree. However for the case with NaCl 20 wt% and the case with the mixed salts in Figure 12, the uncertainty of the dissociation temperature predicted by the model is within 1.7 degree and 1.0 degree respectively.

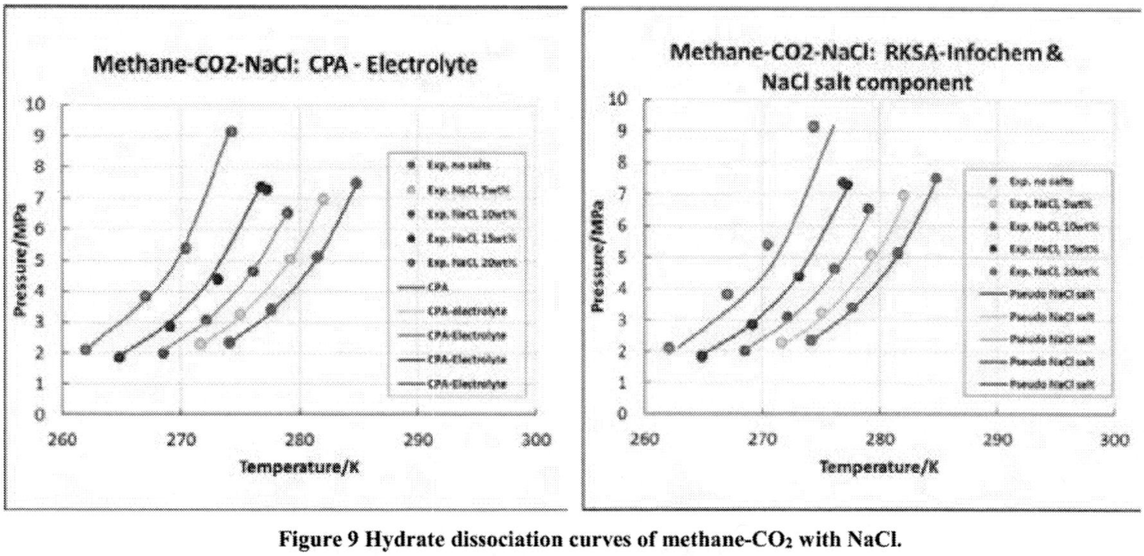

Figure 9 Hydrate dissociation curves of methane-CO_2 with NaCl.

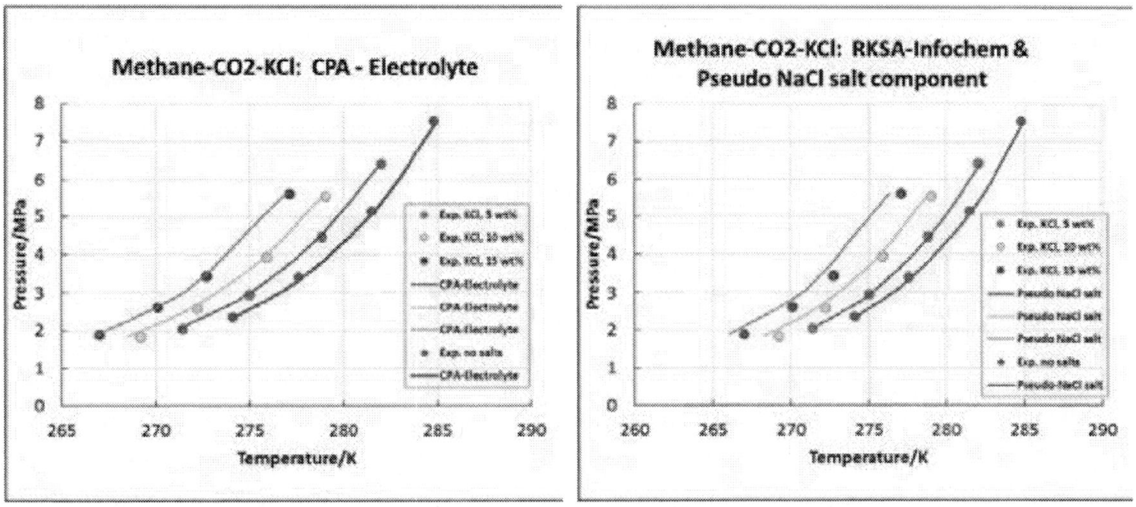

Figure 10 Hydrate dissociation curves of methane-CO_2 with KCl.

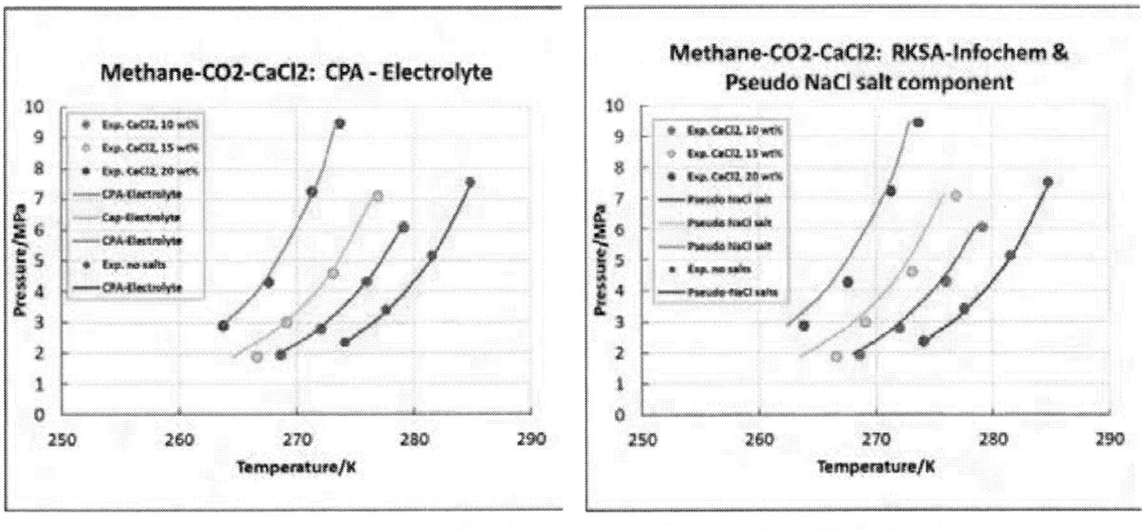

Figure 11 Hydrate dissociation curves of methane-CO_2 with $CaCl_2$.

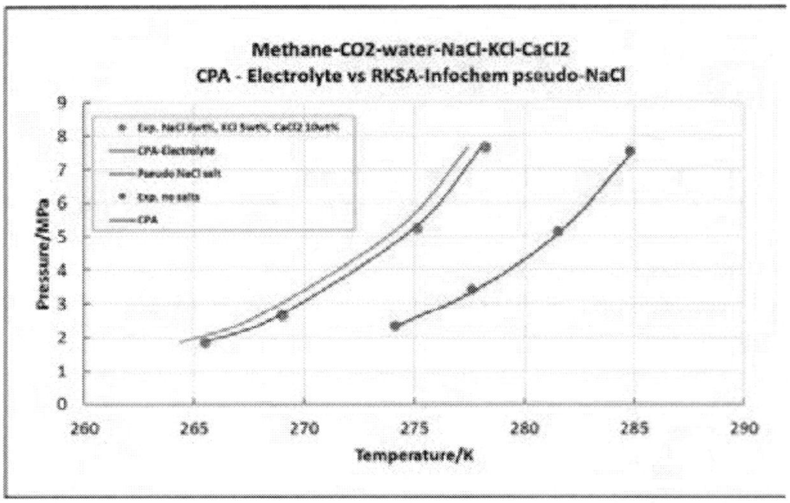

Figure 12 Hydrate dissociation curves of methane-CO_2 with NaCl, KCl and $CaCl_2$.

3.3 Methane-CO_2 with Electrolytes and Methanol

The methane-CO_2 gas mixture with electrolytes and methanol is also applied to validate the two models. The data of hydrate dissociation curves are cited from the work by Dholabhai *et al*[21].

The main objective of this evaluation is to check how the mixed solvents of electrolytes and methanol affect the hydrate formation. Both methanol and the electrolyte concentration of NaCl, KCl and $CaCl_2$ used in the study are 5wt%, 10wt% and 15wt% respectively. One mixed electrolytes of NaCl, KCl and $CaCl_2$ with 5wt% each is also studied. The results of all the cases are summarised in Figure 13 to Figure 15.

As shown in Figure 13 to Figure 15, the hydrate dissociation curves predicted by the CPA-Electrolyte model are in very good agreement with the experimental data for all the cases, whereas the RKSA-Infochem with pseudo-NaCl model works well with lower concentrations of electrolytes and methanol. However with increasing the concentration of the electrolyte and methanol, the hydrate dissociation curves are slightly over-predicted by the model.

Another interesting finding from this study is that the effect of single NaCl and $CaCl_2$ electrolyte on hydrate formation by weight is actually much stronger than that of methanol.

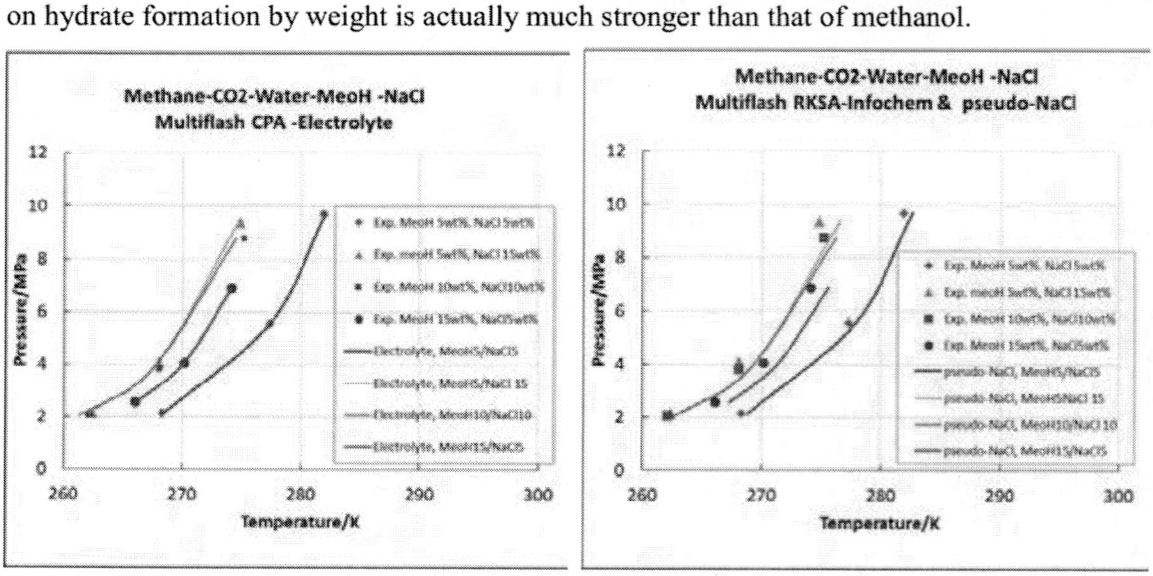

Figure 13 Hydrate dissociation curves of methane-CO_2 with NaCl and methanol.

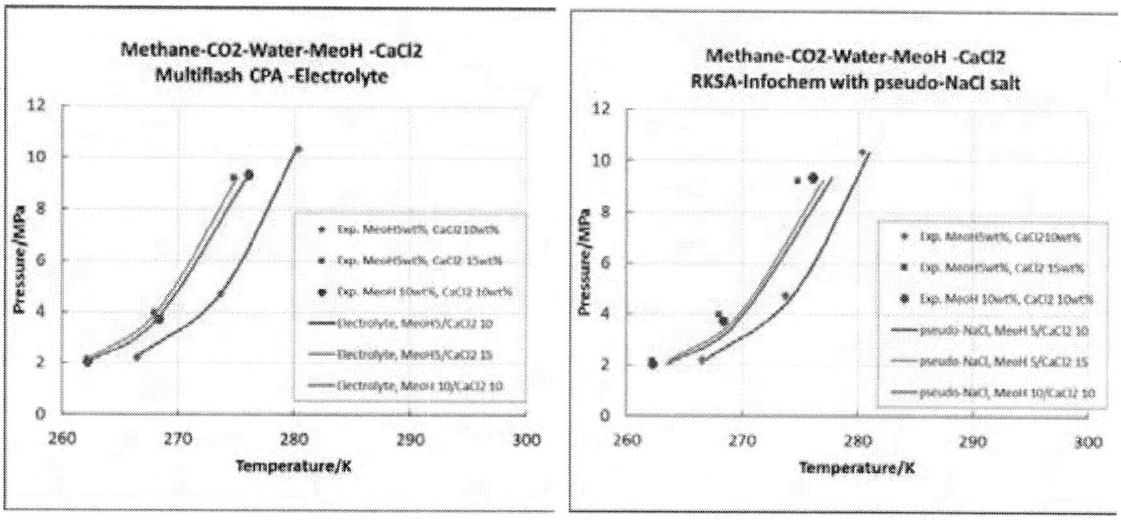

Figure 14 Hydrate dissociation curves of methane-CO_2 with $CaCl_2$ and methanol.

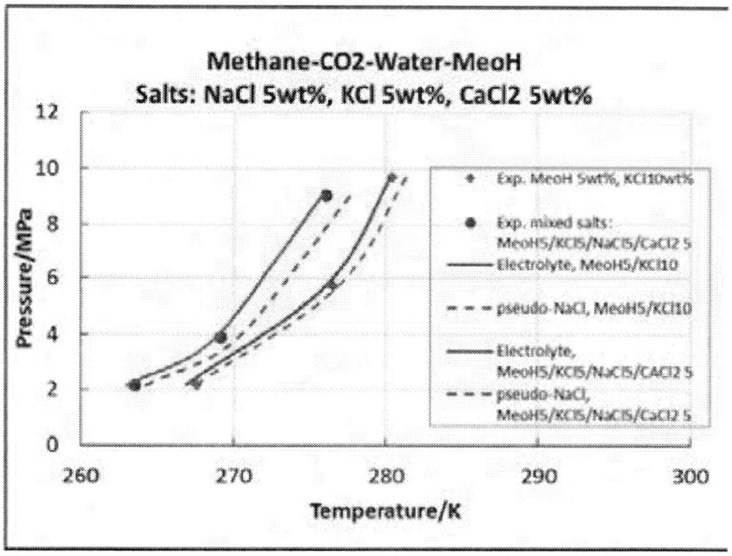

Figure 15 Hydrate dissociation curves of methane-CO_2 with mixed electrolytes and methanol.

3.4 Natural Gas, Gas Condensate and Black Oil with Formation Water

The study on the effect of salts on hydrate formation is now extended to a natural gas, gas condensate and black oil, which are summarised as follows:

- Natural gas with Forties Formation Water (FFW)[22]
- Gas condensate with Synthetic Formation Water (SFW)[23]
- Black oil with mixed electrolytes ($CaCl_2$, $MgCl_2$) and FFW[24]

The three cases are cited from the work by Tohidi *et al*[22-24]. The details of the fluid compositions of the gas and gas condensate are summarised in Table 3, and the composition of the black oil in Table 4. The salt analysis of the Forties formation water and synthetic formation water can be found in Table 2. For mixtures of gas condensate and black oil, a standard fluid characterisation procedure in Multiflash is applied to define all the physical properties of the petroleum fractions required by the thermodynamic fluid models[25].

The hydrate dissociation curves of the natural gas and gas condensate with and without FFW and SFW are presented in Figure 16. The figures clearly show that the predicted hydrate dissociation curves by the two models are in very good agreement with the experimental data. The results also show that the suppression effect by the FFW and SFW on hydrate formation is quite significant, 2 to 3 degree for the natural gas and 2 to 5 degree for the gas condensate, depending on the pressures.

For the black oil, the results are presented in Figure 17, in which the mixed salts A is a mixture of NaCl 6.01wt%, $CaCl_2$ 1.49 wt% and $MgCl_2$ 0.19wt%, whereas the mixed salt B consists of NaCl 8.46 wt%, $CaCl_2$ 3.04wt% and $MgCl_2$ 0.87 wt%. Both models can give accurately prediction on the effect of mixed electrolytes on the hydrate dissociation curves for the black oil. Since the results by the pseudo-NaCl with RKSA-Infochem model is almost the same as that of CPA-Electrolyte, the results are not presented in Figure 17. Based on the results, it is interesting to know that the mixed salt A gives the same effect as the Forties formation water on the hydrate formation.

Components	Natural Gas (mol%)	Gas Condensate (mol%)
Methane	85.93	68.73
Ethane	6.75	9.88
Propane	3.13	4.22
i-Butane	0.71	0.48
n-Butane	0.88	1.49
i-Pentane	-	0.32
n-Pentane	0.57	0.86
C6	-	1.00
C7	-	2.02
C8	-	2.33
C9	-	1.78
C10	-	1.25
C11	-	0.89
C12	-	0.64

C13	-	0.52
C14	-	0.65
C15	-	0.48
C16	-	0.36
C17	-	0.24
C18	-	0.25
C19	-	0.20
C20	-	0.78
CO2	1.31	0.27
N2	0.72	0.36

Table 3 The compositional analysis of the natural gas and gas condensate.

Components	Black oil (mol%)	Molecular weight	SG @60F/60F
METHANE	23.96	16.04	
ETHANE	3.98	30.07	
PROPANE	5.65	44.10	
ISOBUTANE	2	58.12	
N-BUTANE	4.12	58.12	
ISOPENTANE	2.09	72.15	
N-PENTANE	2.89	72.15	
C6	4.1	85	0.666
C7	6.25	94	0.714
C8	6.55	105	0.738
C9	5.2	119	0.761
C10	4.06	133	0.776
C11	3.31	147	0.790

C12	2.64	162	0.801
C13	2.36	173	0.815
C14	2.18	188	0.831
C15	1.95	203	0.837
C16	1.61	220	0.844
C17	1.28	233	0.849
C18	1.27	248	0.856
C19	1.05	262	0.863
C20+	11.5	483	0.947

Table 4 The compositional analysis of the natural gas and gas condensate.

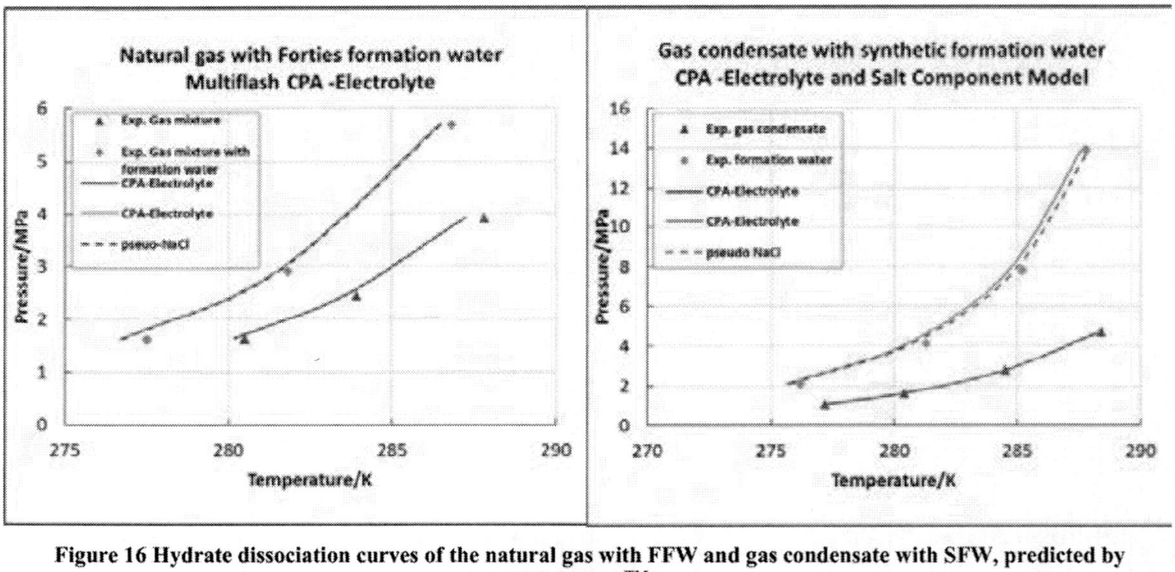

Figure 16 Hydrate dissociation curves of the natural gas with FFW and gas condensate with SFW, predicted by Multiflash^{TM}.

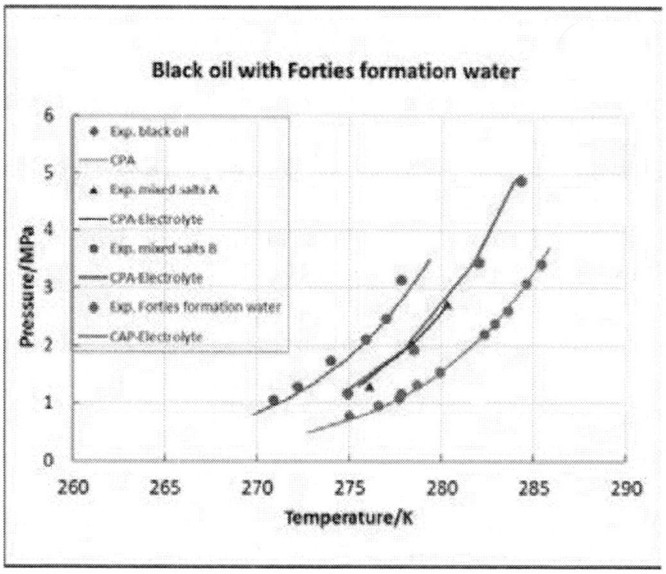

Figure 17 Hydrate dissociation curves of the black oil with mixed electrolytes and FFW, predicted by Multiflash^{TM}.

4 Conclusion

Two well-known thermodynamic models; CPA (Cubic-Plus-Association) with Electrolyte and the advanced version of RKS with modified NRTL type mixing rule were applied to study the suppression effect of single and mixed salts on hydrate formation. The fluid systems used in the study include the pure methane, binary methane-CO_2 gas mixture, natural gas, gas condensate and black oil, and the total salt concentration ranges from 3wt% to 21wt% including Forties formation water and synthetic formation water. For methane-CO_2 gas mixture, the mixed solvents with methanol concentrations of 5wt%, 10wt% and 15wt% are also considered to investigate the effect of mixed solvents on hydrate inhibition.

The study shows that both models are capable of giving accurate predictions on the suppression effect of single and mixed electrolytes on hydrate formation. The average accuracy of the predicted hydrate dissociation temperature is within 0.5 degree for CPA-Electrolyte, and about 1 degree for RKSA-Infochem with pseudo-NaCl salt approach. For the pseudo-NaCl approach, the salts and mixed electrolytes were approximately represented by a single pseudo-NaCl salt component, which works well for the cases with total salt concentration less than 20wt%. However this approach needs to be improved when dealing with mixed solvents involving salts with methanol or glycols and cases with high salt concentrations.

It is very interesting to know that the salts NaCl and $CaCl_2$ by weight are actually stronger than methanol as shown in Figure 13 and Figure 14. This demonstrates that taking into account of the supressing effect of salts in formation and produced water and brines on hydrate formation is important in deciding the correct hydrate inhibitor injection rates in oil and gas production.

In general the CPA-Electrolyte, as a rigorous thermodynamic model, is the recommended model for modelling effects of salts on hydrate inhibition in Multiflash^{TM}. However the RKSA-Infochem with a single pseudo-NaCl salt component has some advantages; simple, robust and is also able to give accurate predictions on the behaviour of fluid and solid phases, having the potential to be used in transient pipeline simulators.

5 References

1. E. Dendy Sloan and Carolyn A. Kon, "*Clathrate Hydrates of Natural Gases*", Third Edition, CRC Press Taylor & Francis Group, 2008.
2. Richard Szczepanski, Beryl Edmonds , Nigel Brown, Terry Hamilton, Technology 1998, 0il & Gas Journal March 9th 1998.
3. B. Edmonds, R. A. S. Moorwood and R. Szczepanski, presented at Deepwater and Ultra Deepwater drilling, London, 17-18 May, 1999.
4. B. Edmonds, R. A. S. Moorwood and R. Szczepanski, Paper presented at SMI Offshore Production Technology Conference, London, 31 January – 1st February, 2001.
5. P.F.Pickering, B. Edmonds, R.A.S.Moorwood, R. Szczepanski, and M.J. Watson, IIR Conference, Aberdeen 16th – 20th September 2001.
6. E.M Hendriks, B. Edmonds, R.A.S. Moorwood and R. Szczepanski, Fluid Phase Equilibria, 117, 193-200, 1996.
7. B. Edmonds, R. A. S. Moonwood, and R. Szczepanski, SPE 35569, 1996.
8. Ripmeester, J. A. and C. I. Ratcliffe, J. Phys. Chem., 94, p8773, 1990.
9. Tohidi, B, A. .Danesh, R.W. Burgass, A.C. Todd, SPE35568, 1990.
10. Ripmeester, J.S. Tse, C.J. Ratcliffe, G.E. Mclaurin, 35, 325 (1987)
11. J.H. van der Waals and J.C. Platteeuw, Adv. Chem. Phys., 2, 1-57, 1959.
12. W.R. Parrish and J.M. Prausnitz, Ind. Eng. Chem. Proc. Des. Dev., 11: 26-35, 1972.
13. Infochem Computer Services Ltd, "*Modelling gas hydrate formation and inhibition for black oils*", Deepstar contract work, November 2002.
14. Georgios M. Kontogeorgis and Georgios K. Folas, *"Thermodynamic Models for Industrial Applications"*, WILEY, A John Wiley and Sons Ltd, Publication, 2010.
15. Peiming Wanga, Kenneth S. Pitzer and John M. Simonson, J. Phys. Chem. Ref. Data, Vol. 27, No. 5, 1998.
16. K. S. Pitzer, "*Activity Coefficients in Electrolyte Solutions*", 2nd Edition (Chemical Rubber, Boca Raton, FL, 1991.
17. R. A. Robirson and R.H. Stokes, "Electrolyte Solutions", 3rd. ed., Butterworth, London, 1970.
18. Hooman Haghighi, Antonin Chapoy, and Bahman Tohidi, Ind. Eng. Chem. Res. 47, 3983–3989, 2008.
19. P. D. Dholabhai, P. Englezos, N. Kalogerakis and P. R. Bishnoi, The Canadian Journal of Chemical Engineering, Vol. 69, June 1991.
20. Pankaj D. Dholabhai and P. Raj Bishnoi, J. Chem. Eng. Data, 39, 191-194, 1994.
21. Pankaj D. Dholabhai, J. Scott Parent, P. Raj Bishnoi, Fluid Phase Equilibria 141, 235-246, 1997.
22. Bahman Tohidi, Ali danesh, R.W. Burgass and A.C. Todd, SPE 28884, 1994.
23. Kasper K., Ostergaard, Bahman Tohidi, Rod W. Burgass, Ali Danesh, and Adrian C. Todd, AIChE Journal, vol. 47, No. 7, July 2001.
24. B. Tohidi, A. Danesh, A. C. Todd and R. W. Burgass, Chemical Engineering Science, Vol. 52, No. 19, pp. 3257 3263, 1997.
25. Multiflash for Windows 6.1, Users' Guide, 2017.

AUTHOR INDEX

AUTHOR INDEX

AIChE
120 Wall Street, FL 23
New York, NY 10005-4020

ISBN 978-1-5108-4150-5

9 781510 841505

CURRAN ASSOCIATES INC.
proceedings.com